FUNGICIDES

CLASSIFICATION, ROLE IN DISEASE MANAGEMENT AND TOXICITY EFFECTS

BIOCHEMISTRY RESEARCH TRENDS

FUNGICIDES

CLASSIFICATION, ROLE IN DISEASE MANAGEMENT AND TOXICITY EFFECTS

MIGUEL N. WHEELER
AND
BRADLEY R. JOHNSTON
EDITORS

nova publishers
New York

For permission to use material from this book please contact us:
Telephone 631-231-7269; Fax 631-231-8175
Web Site: http://www.novapublishers.com

NOTICE TO THE READER

The Publisher has taken reasonable care in the preparation of this book, but makes no expressed or implied warranty of any kind and assumes no responsibility for any errors or omissions. No liability is assumed for incidental or consequential damages in connection with or arising out of information contained in this book. The Publisher shall not be liable for any special, consequential, or exemplary damages resulting, in whole or in part, from the readers' use of, or reliance upon, this material. Any parts of this book based on government reports are so indicated and copyright is claimed for those parts to the extent applicable to compilations of such works.

Independent verification should be sought for any data, advice or recommendations contained in this book. In addition, no responsibility is assumed by the publisher for any injury and/or damage to persons or property arising from any methods, products, instructions, ideas or otherwise contained in this publication.

This publication is designed to provide accurate and authoritative information with regard to the subject matter covered herein. It is sold with the clear understanding that the Publisher is not engaged in rendering legal or any other professional services. If legal or any other expert assistance is required, the services of a competent person should be sought. FROM A DECLARATION OF PARTICIPANTS JOINTLY ADOPTED BY A COMMITTEE OF THE AMERICAN BAR ASSOCIATION AND A COMMITTEE OF PUBLISHERS.

Additional color graphics may be available in the e-book version of this book.

Library of Congress Cataloging-in-Publication Data

ISBN: 978-1-62948-043-5

Library of Congress Control Number: 2013948184

Published by Nova Science Publishers, Inc. † New York

CONTENTS

PREFACE

Fungicides are chemical agents that inhibit or eliminate mycelial growth or fungal spores. The chemical, physical, and biological characteristics of a fungicide determine its suitability for control of a determined disease. In this book, the authors present current research in the study of the classification, role in disease management and toxicity effects of fungicides. Topics discussed in this compilation include plant-derived biofungicides; tricyclazole and azoxystrobin in rice blast management; fungicides and their role in disease management; classification of fungicides; and effective fungicides for cereal crops protection against toxicogenic fungi causing fusarium head blight.

Chapter 1 – Natural products such as the plant-derived ones, e.g. plant extracts and essential oils, as tools of biological control have recently acquired a great scientific and commercial interest. This is due to environmental toxicity problems such as water contamination, persistence of residues in the plant, and to resistance of fungi to chemical fungicides. Many of these plant-derived products are used in the traditional medicine and often they are supposed to be efficient also in plant disease control. In fact, historically, plant products have a determinant role in agriculture and food preservation. The sources of these biofungicides are often plant secondary metabolites, which evolved in plants to confer a protection against microbial attack. Plant-derived biofungicide formulations may be represented by a single active compound, simple mixtures of natural ingredients or complex mixtures with multiple effects on the target pathogens. Such complex mixtures might act also as plant biostimulants. Due to the rich chemical composition of natural substances, it is difficult to identify exactly which component has the antifungal effect. In addition, many compounds act in synergy against different fungal species. The novel active compounds are continuously being discovered, thus responding to the consumer demand on the significant reduction of chemical usage in agriculture. Plant-derived biofungicides are essential to develop new biological control strategies required for the growing organic market. A case study on antifungal activity of *Vitex agnus-castus* extract against *Pythium ultimum* in tomato under *in vitro* and *in vivo* conditions is presented. In addition, the ability of *V. agnus-castus* extract to enhance plant defence responses connected with *PR* gene induction upon pathogen inoculation is discussed.

Chapter 2 – *Magnaporthe oryzae*, the causal agent of the rice blast disease, is one of the most important pathogens of cultivated rice worldwide. It can cause serious epidemics resulting in substantial yield and economic losses. Rice blast management relies on the use of fungicides in many regions. At present, there are approximately thirty fungicides registered worldwide, most of which are being used in Asia. In Europe, azoxystrobin and tricyclazole were registered for application on rice at the beginning of 1990's and in Italy are being extensively used since then. In this chapter, the main chemical classes of fungicides used in rice blast management will be characterized, with emphasis on azoxystrobin and tricyclazole, which will be described in more detail.

The molecular basis of their modes of action and their activity on various stages of pathogen life cycle will be defined. Despite having an important role in rice disease management, only scarce information about responses of the *M. oryzae* isolated from rice to azoxystrobin and tricyclazole are available, therefore the baseline response of the pathogen, risks and possible emergence of resistance to both fungicides will be discussed.

Chapter 3 – Fungal diseases are known to affect produce quality on most crops and to cause severe economic losses worldwide. The use of fungicides has been for a long time the main tool to control many economically important fungal pathogens. In the last decades, increased public concern on food safety and environmental impact of farming activities has led regulatory Authorities to adopt very stringent rules on the evaluation and authorization of plant protection products (PPPs), especially for toxicological and environmental negative side effects related to their use. As a result, the development of new PPPs has been addressed at reducing their toxicity towards humans, mammalians and non-target species. In this framework, agrochemical firms have made available several new fungicides with novel specific (single-site) modes of action. Many of them have penetrant properties, are very selective, showing specific activity against target pathogens, as well as enhanced efficacy as compared to traditional multisite fungicides (i.e., carbamates, copper compounds, dithiocarbamates, quinones, sulphur, thiophtalimides, etc.). On the other hand, their single-site modes of action make them at moderate to high risk of inducing fungicide resistance in target pathogens and, hence, anti-resistance strategies must be carefully implemented in their use. Nowadays, about 200 active substances are available, with almost 40 different known specific target sites, plus a number of chemicals having multi-site or unknown mode of action. The increased availability of fungicides allows a more flexible and effective planning of crop protection strategies well fitting to Integrated Pest Management (IPM) principles. Yet, it requires a constant update of technical and scientific expertise by farmers and advisors. This review focuses on the main features of the most commonly used fungicides and their role in disease control in the overall approach to IPM aiming at improving cropping sustainability.

Chapter 4 – Fungicides are chemicals agents that inhibit or eliminate the mycelial growth or fungal spores. The chemical, physical, and biological characteristics of a fungicide determine its suitability for control of a determined disease. Fungicides can be classified in different ways as fungistactic, anti-sporulant, disinfectant, preventative, protectant, residual, contact, curative, eradicative, topic, deep penetrating, non-penetrating, mesostemic, and systemic, among others. As a result, the literature is confusing to clearly understand and to discuss topics on fungicides. This chapter provides information regarding fungicides classification covering mainly fungicides used in aerial organs of plants.

Chapter 5 – Analyses the long-term effectiveness studies of novel fungicides use for *Fusarium* head blight control on winter wheat in Krasnodar territory of Russia are presented. Effects of fungicides on the grain infestation by the pathogen including latent infection are described. The hazard of studied preparations to the grain agrobiocenosis is shown.

In: Fungicides   ISBN: 978-1-62948-043-5
Editors: M.N. Wheeler, B.R. Johnston   © 2013 Nova Science Publishers, Inc.

Chapter 1

PLANT-DERIVED BIOFUNGICIDES: OVERVIEW AND PERSPECTIVES

Eva Švecová and Paola Crinò[*]
ENEA C.R. Casaccia, Unit of Sustainable Development
and Innovation of the Agro-Industrial System, Roma, Italy

ABSTRACT

Natural products such as the plant-derived ones, e.g. plant extracts and essential oils, as tools of biological control have recently acquired a great scientific and commercial interest. This is due to environmental toxicity problems such as water contamination, persistence of residues in the plant, and to resistance of fungi to chemical fungicides. Many of these plant-derived products are used in the traditional medicine and often they are supposed to be efficient also in plant disease control. In fact, historically, plant products have a determinant role in agriculture and food preservation. The sources of these biofungicides are often plant secondary metabolites, which evolved in plants to confer a protection against microbial attack. Plant-derived biofungicide formulations may be represented by a single active compound, simple mixtures of natural ingredients or complex mixtures with multiple effects on the target pathogens. Such complex mixtures might act also as plant biostimulants. Due to the rich chemical composition of natural substances, it is difficult to identify exactly which component has the antifungal effect. In addition,

[*] corresponding author: paola.crino@enea.it.

many compounds act in synergy against different fungal species. The novel active compounds are continuously being discovered, thus responding to the consumer demand on the significant reduction of chemical usage in agriculture. Plant-derived biofungicides are essential to develop new biological control strategies required for the growing organic market. A case study on antifungal activity of *Vitex agnus-castus* extract against *Pythium ultimum* in tomato under *in vitro* and *in vivo* conditions is presented. In addition, the ability of *V. agnus-castus* extract to enhance plant defence responses connected with *PR* gene induction upon pathogen inoculation is discussed.

Keywords: Plant extract, *PR* gene activation, Disease control, Sustainable agriculture

1. INTRODUCTION

Fungal pathogens represent serious threats for vegetable crops under field and protected conditions. In the past, losses of crop yield caused by fungal diseases have had severe effects even on the human race leading to famines. Therefore, the proper management of plant fungal diseases is of global importance. According to Agrios (2005), almost all control methods protect plants from becoming diseased but only some plant diseases can be well controlled by therapeutic means under field conditions. Therefore, new crop protection strategies are continuously being investigated.

In the past, excessive use of agrochemicals has led to considerable changes in people's attitudes towards the use of pesticides in agriculture (Pal and McSpadden Gardener, 2006). Nowadays, consumer perceptions worldwide are that chemical usage in agricultural production needs to be significantly reduced. Natural preparations may have many advantages for both grower and consumer, e.g. in the crop culture, the environment is less stressed by natural products and, in addition, fungal resistance against such products has not been acquired. In this way, natural products are often preferred by consumers as they are considered healthier due to the minor content of chemical residues. However, growers still often use intensively chemical fungicides because they can significantly contribute to the improvement in crop productivity and quality. As even today many of the fungal diseases are difficult to control, synthetic fungicides are still the most important means for their management. More than 100 fungicides have been developed, and hundreds of fungicide formulations are available (Vidhyasekaran, 2004).

In the past decades, various disadvantages of a wide use of chemical fungicides have emerged. Toxicity problems such as water contamination, persistence of residues both in the environment and in the food or accumulation in animal and human fat tissues have been reported (Butt et al., 2001). In some cases, other effective methods of control are not available and chemical pesticides are banned or being phased out because they are harmful for the environment (Li et al., 2008). Resistance of fungi to some systemic fungicides, all of which containing a benzene ring, first appeared in the 1960s (Agrios, 2005). To date, as reported by Agrios (2005) and Vidhyasekaran (2004), fungicide resistance has been found in *Phytophthora infestans, Peronospora parasitica,* and *Bremia lactucae* against phenylamides (metalaxyl compounds); in *Ustilago nuda* and *Ustilago maydis* against carboxamides (carboxin compounds); in *Erysiphe graminis* f. sp. *hordei* against hydroxypyrimidine; in *E. graminis* f. sp. *tritici, E. graminis* f. sp. *hordei,* and *Pyrenophora teres* against triazoles (propiconazole); in *Botrytis cinerea, Botrytis fabae, Septoria tritici, Pyrenopeziza brassicae, Cercospora beticola, Fusarium nivale, Fusarium culmorum,* and *Pseudocercosporella herpotrichoides* against benzimidazoles (carbendazim, benomyl, thiophanate-methyl compounds); and also in some strains of other important pathogens such as *Alternaria, Colletotrichum, Verticillium, Sphaerotheca, Mycosphaerella, Aspergillus, Penicillium, Pythium,* and *Venturia inaequalis.*

Biological control strategies, especially for the growing organic market, are urgently required (Butt et al., 2001). In particular, use of plant-derived biofungicides is a promising management strategy for a safer and healthier environment (Peluola et al., 2010; Shafique et al., 2011). Biofungicides of plant origin are defined as naturally occurring substances that control diseases (biochemical pesticides), and are approved for organic production (Francis and Keinath, 2010). They can be a viable alternative to chemical fungicides and can be used as part of an integrated disease management program to reduce the risk of pathogens developing resistance to chemical fungicides. Therefore, plant-derived products, as a method of biological control, have recently acquired a great scientific and commercial interest as they are a natural source of antifungal compounds.

The major advantages of using the natural fungicides consist in:

- reducing the use of chemical fungicides;
- their possible use by organic growers;

- broad-spectrum of activity against numerous fungal pathogens found in some plant-derived-products;
- existence in nature as preformed defences with activity retained in diverse environmental conditions;
- reduced phytotoxicity compared to synthetic fungicides, even if reapplied several times (Thomas, 2009).

However, some disadvantages in using biofungicides exist, e.g. narrow target range of diseases to control; slow mode of action, while a plant infection needs to be eradicated fast; shorter shelf life than chemical controls.

2. THE ORIGIN OF PLANT-DERIVED BIOFUNGICIDES

Historically, plant products play an important role in the fields of agriculture and food preservation. It is also reported that about 60 percent of the essential oils obtained from plants possess antifungal activity (Suresh et al., 1997).

The term "biological control" comprises the use of microbial antagonists and various natural products extracted or fermented from various sources, including plant-derived products. Plant-derived products may be represented by simple crude extracts or their mixtures, by single active compounds or their complex mixtures with multiple effects on the plant as well as on the pathogen.

Plants produce an array of over 100 000 low-molecular-mass compounds called secondary metabolites (Dixon, 2001). Secondary metabolites are produced within the plants besides the primary biosynthetic and metabolic pathways of compounds aimed at plant growth and development, such as carbohydrates, amino acids, proteins and lipids (Bernhoft, 2010). Although, they are not generally essential for the basic metabolic processes (Dixon, 2001), several of them are found to hold important functions in the living plants, e.g. flavonoids can protect the plant against free radicals generated during photosynthesis (Bernhoft, 2010). Many secondary metabolites provide chemical barriers against animals and microorganisms. These chemicals may have general or specific activity against key target sites in bacteria, fungi and viruses (Martínez, 2012).

Most of the naturally active substances show fungistatic as well as fungitoxic characteristics depending on their concentration. Due to their rich chemical composition, it was often difficult to identify exactly which component has the antifungal effect; this is now possible using advanced techniques allowing selective identification of the antifungal compounds. Numerous plant compounds have similar fungicidal characteristics, so the spectrum of their activity against different fungal species can be considered as a synergistic effect (Pepeljnjak et al., 2003).

3. MODE OF ACTION AND CLASSIFICATION

The mode of action of the plant-derived biofungicides can be either direct or indirect. The direct mode of action consists in the direct affecting plant pathogen. The plant-derived product can be harmful to the pathogen at the direct contact and/or can cause changes in the environment surrounding the pathogen in such a way to create unsuitable conditions for pathogen's development. Indirect mode of action is characteristic for the plant-derived products that are not directly harmful to the pathogen, but activate the defence responses in plant. The plant-derived products can act even by both modes of action (Švecová et al., 2013). In addition, their mode of action might be influenced by the concentration applied.

Generally, fungal pathogens respond in three different ways to treatment with plant metabolites: inhibitory, stimulatory or with lack of any effect (Bailey, 2010). Inhibitory effect occurs when presence of plant metabolites leads to a growth delay or complete death of fungal pathogens. Stimulatory effect leads to an increase of fungal growth in comparison with the untreated control, while no effect is a similar response in growth as in untreated pathogen.

Plant-derived fungicides might be divided into three categories as proposed by Martínez (2012):

- Preformed compounds (preformed resistance in plants)

Phytonzides are all the substances with antimicrobial properties naturally occurring in plants under their biologically active form. The preformed compounds are contained in extracts and essential oils obtained from plants. They can be phenolic compounds, hydroxycinnamic acids, flavonoids, plant growth substances and regulators, glucosinolates, latex and steroids, acetaldehyde and other compounds.

- Inducible preformed compounds (inducible preformed resistance in plants)

Some substances, which are normally present in healthy tissues, can be further induced in the host in response to pathogen attack, as well as to other stresses. There are also compounds that occur as inactive precursors and are activated in response to pathogen attack. These preformed compounds (phytoanticipins) differ from the inducible phytoalexins that are synthesized from remote precursors in response to pathogen attack (Prusky, 1997).

- Induced inhibitory compounds

Inducible resistance mechanisms are active, energy-requiring systems typified by specific recognition of an invader that ultimately leads to the production of proteins or metabolites antagonist to the invader. Among these inhibitory compounds, phytoalexins, pathogenesis-related proteins (PR), active oxygen species and lectins are included.

Several examples of plant-derived products controlling diseases by activating signal transduction systems have been reported. Milsana, a commercial product from *Reynoutria sachalinensis,* significantly reduced incidence of powdery mildew *(Podosphaera xanthii)* in cucumber (Wurms et al., 1999; Daayf et al., 2000) increasing the levels of hydroxycinnamic acid, *p*-coumaric, caffeic, ferulic acids and *p*-coumaric acid methyl ester in the cucumber leaves. All these phenolic compounds showed antifungal activity against common pathogens of cucumber, such as *B. cinerea, Pythium ultimum,* and *Pythium aphanidermatum.* This suggests that this plant extract would induce synthesis of antifungal compounds, so contributing to disease resistance (Daayf et al., 2000). Also pokeweed (*Phytolacca* spp.) antiviral protein (PAP) induces synthesis of PR proteins along with a small increase in salicylic acid levels in tobacco (Smirnov et al., 1997); the *Yucca schidigera* extract, showing similar efficiency as resistance inducer [acibenzolar-S-methyl (ASM)] to control *Venturia inequalis*, up-regulated two genes encoding PR proteins in apple (Bengtsson et al., 2009); systemic acquired resistance provided by *Datura metel* extract against downey mildew disease caused by *Sclerospora graminicola* in pearl millet was demonstrated (Devaiah et al., 2009). Moreover, continuous advances in molecular technologies contribute to understand the complex pathways of plant natural product biosynthesis. Genetic and reverse genetic approaches, together with metabolic engineering of natural product pathways, show to be a feasible strategy for enhancement of plant disease resistance (Dixon, 2001).

4. PLANT-DERIVED PRODUCTS WITH KNOWN ANTIFUNGAL PROPERTIES

Thousands of plant species are known to have medicinal properties due to their chemical content. Plant-derived products such as plant extracts, essential oils and powdered plant material are traditionally used in local medicine all over the world. Therefore, these products are supposed to have antimicrobial properties in plant disease management as well.

Recently, many studies have appeared describing antifungal compounds isolated from various plants and tested *in vitro* and/or *in vivo* (Table 1). However, only few of these compounds have been tested for crop-disease management under field conditions. For example, the aqueous extracts of the leaves of *Carica papaya* and *Vernonia amygdalina* revealed that, under the field conditions, they can be used as biofungicides against foliar fungal pathogens of groundnut (*Arachis hypogaea* L.) such as cercospora leaf spots and rust, web blight caused by *Phoma arachidicola*, Botrytis blight (*B. cinerea*) and pepper spot (*Leptosphearrulina crassiaca*; Ogwulumba et al., 2008). An efficiency of a plant-derived product is influenced by many factors such as extraction methods, concentration, part of plant used, etc. Some studies demonstrated that the material from the same plant shows different efficiency depending on the form in which the material is applied or on the extraction method, e.g. powder application of *Pithecellobium dulce* on strawberry fruit compared with aqueous extracts had a better fungicidal effect against *B. cinerea*, *P. digitatum*, and *R. stolonifer*, since powders are not subjected to an extraction process. It is supposed that powders may contain also various active compounds that might be acting all together (Bautista-Baños et al., 2003); methanolic extracts of *B. racemosa* L. showed excellent inhibitory activity against *Fusarium* sp., *Trichoderma kongii*, *Penicillium* sp., *Ganoderma tropicum*, *G. lucidum*, *Aspergillus* sp. and *Rhizopus* sp. in comparison with ethanolic and boiling water extracts (Hussin et al., 2009); among different solvent extracts of *D. hamiltonii* (Wight&Arn), petroleum ether extracts showed the highest antifungal activity against important phytopathogenic fungi causing diseases in sorghum, maize and paddy such as *Fusarium* spp., *Drechslera* spp., *Aspergilus* spp, *Penicillium* spp. and *A.*

alternata (Mohana et al., 2008); the phenolic fractions of the *R. ulmifolius* extract, rich in tannins, resulted active against *Alternaria* sp., *Aspergillus* spp., *Botrytis* sp., *Drechslera* sp., *Fusarium* sp., *Penicillium* sp., *Rhizopus* sp., *Trichoderma* sp., *Ulocladium* sp. and *Verticillium* sp., whereas the fractions containing chlorogenic, caftaric acid, and caffeoyl derivates resulted partially effective (Sisti et al., 2008). The extracts of different parts of the same plant may have different degrees of inhibition. In the study of Mdee et al. (2009) leaf extracts were more active than seed or flower extracts. In some cases, a plant-derived product that at a certain concentration inhibits the pathogen, at different concentration may promote its growth, e.g. the extracts of *Ricinus communis* at higher concentration acted as growth promoter to *C. destructivum*, since its mycelium had a better sporulation and fluffiness than that of the control (Akinbode and Ikotun, 2008). The antifungal activity of plant-derived products varies also according to an active compound and target species, e.g. three antifungal lignans isolated from methanol extract of nuttmeg tree [*Myristica fragrans* (Houttyn)] showed different activity against *A. alternata, C. coccodes, C. gleosporoides* and *M. grisea* (Cho et al., 2007).

In order to obtain the most effective biofungicide, the researchers aim at identifying an active compound that is responsible for the fungicidal effect, e.g. kaempferol isolated from *P. dulce* (Bautista-Baños et al., 2003), quinolone alcaloid from *R. graveolens* (Oliva et al., 2003), fistulosin from roots of *A. fistulosum* (Phay et al., 1999), nyasol from *A. aphodeloides* (Park et al., 2003), phenolic acids and flavonoids from *B. racemosa* (Hussin et al., 2009).

Some of the plant-derived products were reported to have activity, which is comparable with that of chemical fungicides or even more effective, e.g. garlic and grapefruit juice (BioSept) effectively controlling rose powdery mildew (*Sphaerotheca pannosa* var. *rosae*) has an efficacy equal to that of the standard fungicide triforine (Wojdyla, 2000); active compound of *D. hamiltonii* resulted more efficient than synthetic fungicides in controlling the growth of *Penicilium* spp. (Mohana et al., 2008); the mycelial inhibition by essential oils of three *Cymbopogon* spp. on *Fusarium solani, F. moniliforme, C. lunata, E. rostratum, P. sorgina, C. graminicola, B. sorokiniana* and *Acremonium strictum* were similar or better than those obtained with the chemical control (Zida et al., 2008); *D. kilimandscharicus* showed a broad-spectrum antifungal activity inhibiting mycelial growth of *B. cinerea, B. dothidea* and *P. ultimum* in a way comparable with synthetic fungicides (Tegegne and Pretorius, 2007); *M. oleifera* was compared favourably with the systemic fungicide benomyl in the control of the *Colleotrichum destructivum* on cowpea seeds (Akinbode and Ikotun, 2008).

Table 1. Plants producing substances with antifungal properties

Plant species	Pathogen controlled	Reference
Abrus precatorius L.	*Colleotrichum capsici, Alternaria alternata*	Anand and Bhaskaran (2009)
Acacia nilotica (L.) Willd. ex Delile	*Fusarium* spp.	Satish et al. (2009)
Acacia plumosa Lowe	*Aspergillus niger, Thielaviopsis paradoxa, Colleotrichum* sp.	Lopes et al. (2009)
Aegle marmelos (L.) Corr.Serr.	*Colleotrichum capsici, A. alternata*	Anand and Bhaskaran (2009)
Achras zapota L.	*Fusarium* spp.	Satish et al. (2009)
Ailanthus excelsa (Roxb.)	*Lasiodiplodia theobromae, Fusarium oxysporum*	Ramakrishna Mission (2008)
Allium cepa L.	*A. alternata, Colleotrichum gleosporioides*	Ramakrishna Mission (2008)
Allium fistulosum L.	*F. oxysporum*	Phay et al. (1999)
Allium obliquum L.	*A. niger, B. cinerea, F. oxysporum* f. sp. *gladioli, Penicillium expansum and Sclerotinia sclerotiorum*	Pârvu et al. (2010)
Allium sativum L.	*Phytophthora capsici*	Demirci and Dolar (2006)
	Alternaria brassisicola, B. cinerea, Plectosphaerella cucumerina, Magnaporthe grisea, P. infestans	Curtis et al. (2004)
	F. oxysporum	Ramakrishna Mission (2008)
	A. alternata, Fusarium sp.	Sharma et al. (2010a)
Aloe barbadensis Miller	*F. oxysporum*	Taiga et al. (2008)
Aloe vera L.	*F. oxysporum, A. niger, Rhizopus stolonifer, Penicilium oxalicum*	Taiga and Oufolaji (2008)
Annemarrhena aphodeloides Bunge	*M. grisea, Rhizoctonia solani, P. capsici*	Park et al. (2003)

Table 1. (Continued)

Plant species	Pathogen controlled	Reference
Annona cherimola Mill.	*R. stolonifer*	Bautista-Baños et al. (2000)
Annona muricata L.	*Colleotrichum destructivum*	Akinbode and Ikotun (2008)
Apium graveolens L.	*R. solani, F. oxysporum f. sp. vasinfecum*	Liu et al. (2012)
Argemone mexicana L.	*F. solani, R. solani, Macrophomina phaseolina*	Siddiqui et al. (2002)
Arnica longifolia DC. Eat.	*Colleotrichum acutatum, Colleotrichum fragariae, Colleotrichum gleosporoides*	Tabanca et al. (2006, 2007)
Artemisia nilagirica (C.B.Clarke) Pamp.	*R. solani, Sclerotium rolfsii, M. phaseolina*	Sati et al. (2013)
Aster hesperius A. Gray	*C. acutatum, C. fragariae, C. gleosporoides*	Tabanca et al. (2006, 2007)
Azadirachta indica A. Juss	*Fusarium moniliforme*	Wokoma and Nwaejike (2008)
	F. oxysporum	Taiga et al. (2008)
	F. oxysporum, A. niger, R. stolonifer, P. oxalicum	Taiga and Oufolaji (2008)
Barringtonia racemosa L.	*Fusarium* sp., *Trichoderma kongii, Penicillium* sp., *Ganoderma tropicum, Ganoderma lucidum, Aspergillus* sp., *Rhizopus* sp.	Hussin et al. (2009)
Bazzania trilobata (L.) S.F. Gray	*B. cinerea, Cladosporium cucumerinum, P. infestans, Pyricularia oryzae, Septoria tritici*	Scher et al. (2004)
Brassica juncea L. Czern.	*F. solani*	Sharma et al. (2010b)

Plant species	Pathogen controlled	Reference
Brassica oleracea L.	*P. capsici*	Demirci and Solar (2006)
Bromelia hemisphaerica Lamarck	*R. stolonifer*	Bautista-Baños et al. (2000)
Calotropis procera (Ait)	*F. oxysporum, L. theobromae*	Ramakrishna Mission (2008)
Campuloclinium macrocephalum (Less.) DC.	*C. gleosporoides, F. oxysporum*	Mdee et al. (2009)
Capsicum spp. L.	*B. cinerea*	Wilson et al. (1997)
Capsicum annuum L.	*Phomopsis azadirachtae*	Fathima et al. (2009)
Carica papaya L.	*R. stolonifer*	Bautista-Baños et al. (2000)
	Phoma arachidicola, B. cinerea, Leptosphearrulina crassiaca	Ogwulumba et al. (2008)
	F. oxysporum	Taiga et al. (2008)
	F. oxysporum, A. niger, R. stolonifer, P. oxalicum	Taiga and Oufolaji (2008)
Carthamus oxyacantha M. Bieb.	*F. oxysporum, R. solani*	Abdolmaleki et al. (2011a)
Catalpa ovata G. Don.	*M. grisea, B. cinerea, P. infestans, Puccinia recondita, Blumeria graminis* f.sp. *hordei*	Cho et al. (2006)
Catharanthus roseus L.	*A. alternata, F. oxysporum*	Ramakrishna Mission (2008)
Cromolaena odorata (L.)	*A. alternata, Alternaria longipes*	Ramakrishna Mission (2008)
Cinnamomum zeylanicum J.Presl	*B. cinerea*	Wilson et al. (1997)
Cryptomeria japonica (L. f.) D. Don	*P. capsici, F. oxysporum, Pythium splendens*	Hwang et al. (2005)

Table 1. (Continued)

Plant species	Pathogen controlled	Reference
Cupressus benthamii (Endl.) D.P. Little	*P.infestans*	Goufo et al. (2008)
Curcuma domestica Val.	*P. azadirachtae*	Fathima et al. (2009)
Cymbopogon citratus (DC) Stapf	*F. oxysporum*	Ramakrishna Mission (2008)
Cymbopogon martinii (Roxb.) Wats.	*B. cinerea*	Wilson et al. (1997)
	P. oryzae, Drechslera oryzae, R. solani, Colleotrichum lindemuthianum, Colleotrichum capsici, M. phaseolina, A. alternata, Phylostica sp., *Pestalotia theae, Curvularia lunata, F. oxysporum, Granaria uvicola, S. sclerotiorum, Mysosphaerella* sp., *A. niger, Aspergillus flavus, Botryodiplodia theobromae, Cercospora nicotiana, Phoma* sp.	Sridhar et al. (2003)
Cymbopogon spp. Spreng.	*F. solani, F. moniliforme, C. lunata, Exserohilum rostratum, Phoma sorgina, Colleotrichum graminicola, Bipolaris sorokiniana, Acremonium strictum*	Zida et al. (2008)
Datura metel L.	*Sclerospora graminicola*	Devaiah et al. (2009)
Datura stramonium L.	*Fusarium* spp.	Satish et al. (2009)
	F. oxysporum	Ramakrishna Mission (2008)

Plant species	Pathogen controlled	Reference
Decalepis hamiltonii (Wight&Arn)	*Fusarium* spp., *Drechslera* spp., *Aspergilus* spp., *Penicillium* spp., *A. alternata*	Mohana et al. (2008)
Desmos chinensis Lour.	*Bipolaris oryzae, P. oryzae, Sclerotium rolfsii*	Plodpai et al. (2013a)
	R. solani	Plodpai et al. (2013b)
Deterium microcarpum (Tallow)	*F. oxysporum, A. niger, Penicillium digitatum*	Doughari e Nuya (2008)
Diospyros lotus L.	*F. oxysporum*	Ramakrishna Mission (2008)
Dolichos kilimandscharicus Harms ex Taub.	*B. cinerea, B. dothidea, P. ultimum*	Tegegne and Pretorius (2007)
Emblica officinalis L.	*Fusarium* spp.	Satish et al. (2009)
Eucalyptus sp. L'Hér.	*F. oxysporum*	Abdolmaleki et al. (2011b)
Eucalyptus citridora Hook	*P. azadirachtae*	Fathima et al. (2009)
Eucalyptus globules Labill.	*Fusarium* spp.	Satish et al. (2009)
Eugenia caryophyllata Thunb.	*B. cinerea*	Wilson et al. (1997)
Foeniculum vulgare L.	*P. azadirachtae*	Fathima et al. (2009)
Halodule wrightii Ascherson	*Fusarium* spp.	Ross et al. (2008)
Chrysothamnus nauseosus (Pall. ex Pursh) Britton	*C. acutatum, C. fragariae, C. gleosporoides*	Tabanca et al. (2006, 2007)
Lantana camara L.	*F. oxysporum*	Mdee et al. (2009)
	P. aphanidermatum	Mostafa et al. (2012)

Table 1. (Continued)

Plant species	Pathogen controlled	Reference
Larrea divaricata Cav.	*Fusarium graminearum*	Vogt et al. (2013)
Lysimachia foenum-graecum Hance	*B. graminis* f.sp. *hordei, Puccinia recondita, M. grisea*	Park et al. (2008)
Lavandula sp. L.	*P. capsici, P. megakarya, P. palmivora*	Widmer and Laurent (2006)
Lawsonia inermis L.	*Fusarium* spp.	Satish et al. (2009)
Lupinus mexicanus Cerv.	*R. solani, Sclerotium rolfsii*	Zamora-Natera et al. (2008)
Macaranga monandra Müll.Arg.	*C. aculatum, C. fragariae, C. gleosporoides, F. oxysporum, B. cinerea, Phomopsis obscurans, Plasmopara viticola*	Salah et al. (2003)
Marchantia polymorpha L.	*A. flavus, A. niger*	Mewari and Kumar (2008)
Medicago sativa L.	*P. capsici*	Demirci and Solar (2006)
Mentha piperita L.	*R. solani, F. oxysporum, Phytophthora dershleri, Bipolaris sorkiniana*	Abdolmaleki et al. (2011b)
Mimusops elengi L.	*Fusarium* spp.	Satish et al. (2009)
Moringa oleifera Lam	*Colleotrichum destructivum*	Akinbode and Ikotun (2008)
Myristica fragrans (Houttyn)	*A. alternata, Colleotrichum coccodes, C. gleosporoides, M. grisea*	Cho et al. (2007)
	P. azadirachtae	Fathima et al. (2009)
Nicotiana tabacum L.	*F. oxysporum*	Taiga et al. (2008)
	F. oxysporum, A. niger, R. stolonifer, P.oxalicum	Taiga and Oufolaji (2008)
Ocimum gratissimum L.,	*Botryosphaeria rhodina, Rhizoctonia* sp., *Alternaria* sp.	Faria et al. (2006)

Plant species	Pathogen controlled	Reference
Ocimum sanctum L.	*A. alternata, Fusarium sp.*	Sharma et al. (2010a)
Origanum syriacum Sieb. Exs. Et. L.	*B. cinerea, Alternaria solani, Penicillium* sp., *Cladosporium* sp., *F. oxysporum* f.sp. *melonis, Verticillium dahliae*	Abou-Jawdah et al. (2002)
Orthosiphon stamineus Benth.	*B. cinerea, R. solani, F. solani, Colleotrichum capsici, P. capsici*	Hossain et al. (2008)
Peltophorum pterocarpum (DC.) K. Heyne	*Fusarium* spp.	Satish et al. (2009)
Philonotis revoluta Bosch & Lac.	*F. moniliforme, Helminthosporium turcicum , C. lunata*	Deora et al. (2010)
Pinus roxburghii Sarg.	*Botrytis elliptica*	Sharma and Dhancholia (2008)
Piper caldense C. DC.	*Cladosporium cladosporoides, Cladosporium sphaerospermum*	Freitas et al. (2009)
Piper nigrum L.	*B. graminis* f.sp. *hordei, Puccinia recondita, M. grisea*	Park et al. (2008)
Pithecellobium dulce (Roxb.) Benth.	*B. cinerea, P. digitatum, R. stolonifer*	Bautista-Baños et al. (2003)
Polyathia longifolia (Sonner)	*Fusarium* spp.	Satish et al. (2009)
	A. alternata, A. dianthicola, A. longipes, F. oxysporum	Ramakrishna Mission (2008)
Pongamia pinnata(L.) Pierre	*A. flavus*	Srivastava et al. (1997a, b)
Prosopis juliflora (Sw.) DC.	*Fusarium* spp.	Satish et al. (2009)
Psidium guajava L.	*A. alternata, F. oxysporum*	Ramakrishna Mission (2008)
Punica granatum L.	*Fusarium* spp.	Satish et al. (2009)

Table 1. (Continued)

Plant species	Pathogen controlled	Reference
Ranunculus sp.	*B. elliptica*	Sharma and Dhancholia (2008)
Rosa chinensis Jacq.	*A. alternata*	Ramakrishna Mission (2008)
Rosmarinus sp.	*P. capsici, Phytophthora megakarya, Phytophthora palmivora*	Widmer and Laurent (2006)
Rubus ulmifolius Schott	*Alternaria* sp., *Aspergillus* spp., *Botrytis* sp., *Drechslera* sp., *Fusarium* sp., *Penicillium* sp., *Rhizopus* sp., *Trichoderma* sp., *Ulocladium* sp., *Verticillium* sp.	Sisti et al. (2008)
Ruppia maritima L.	*Fusarium* spp.	Brooker et al. (2008)
Ruta graveolens L.	*B. cinerea, Phomopsis* spp.	Oliva et al. (2003)
Santalum album L.	*B. graminis* f.sp. *hordei, Puccinia recondita, M. grisea*	Park et al. (2008)
Scaligeria tripartita (Kalen.) Tamamsch	*C. acutatum, C. fragariae, C. gleosporoides*	Tabanca et al. (2006, 2007)
Siparuna guianensis Aubl.	*B. cinerea, Ascochyta rabiei*	Švecová et al. (2010)
Solanum lycopersicum L., *Solanum chmielewskii* (C.M. Rick, Kesicki, Fobes & M. Holle) D.M. Spooner, G.J. Anderson & R.K. Jansen	*A. alternata, B. cynerea, F. oxysporum, F. sambucinum, F. solani, F. semitectum, Pyrenochaeta lycopersici, R. solani, S. rolfsii, Sclerotinia sp., Stemphylium sp.*	Zaccardelli et al. (2011)
Syzygium cumini (L.) Skeels.	*Fusarium* spp.	Satish et al. (2009)
Syringodium filiforme Kuetz	*Fusarium* spp.	Ross et al. (2008)

Plant species	Pathogen controlled	Reference
Tephrosia apollinea L.	*A. alternata, Helminthosporium sp., C. acutatum, Pestalotiopsis sp*	Ammar et al. (2013)
Thalassia testudinum Banks ex König	*Fusarium* spp.	Ross et al. (2008)
Thymus vulgaris L.	*R. stolonifer*	Bhaskara Reddy et al. (1997)
	F. oxysporum, P. aphanidermatum, R. solani	Mostafa et al. (2012)
Thymus zygis Loefl. ex L.	*B. cinerea*	Wilson et al. (1997)
Tridax precumbens L.	*F. oxysporum*	Taiga et al. (2008)
	F. oxysporum, A. niger, R. stolonifer, P.oxalicum	Taiga and Oufolaji (2008)
Vernonia amygdalina Delile	*P. arachidicola, B. cinerea, Leptosphearrulina crassiaca*	Ogwulumba et al. (2008)
	C. destructivum	Akinbode and Ikotun (2008)
Vetiveria zizanioides (L.) Roberty	*P. infestans*	Goufo et al. (2008)
Vincetoxicum rossicum (Kleopow) Barbar.	*Fusarium* spp., *B. cinerea, A. alternata, S. sclerotiorum, V. dahliae*	Mogg et al. (2008)
Zingiber officinale Roscoe	*A. alternata, Fusarium* sp.	Sharma et al. (2010a)
	F. oxysporum, P. aphanidermatum, R. solani	Mostafa et al. (2012)

However, in the study of Anand and Bhaskaran (2009), the extract of *Abrus precatorius* was effective in the pot culture experiment against *Colleotrichum capsici* and *Alternaria alternata*, but it was not as efficient as the synthetic fungicide (carbendazim 0.1%).

Recently, many commercial natural products are being developed. Commercial formulations of various plant extracts, such as pepper/mustard (chilli pepper extract and the essential oil of mustard), cassia (extract of cassia tree), clover (70 percent clove oil), and neem (90 percent neem oil) showed their efficacy to control muskmelon wilt caused by *F. oxysporum* f. sp. *melonis* (Bowers and Locke, 2000); Milsana®, an extract from *R. sachalinensis,* confers resistance to powdery mildew and other diseases (Wurms et al., 1999; Daayf et al., 2000).

5. SCREENING METHODS OF ANTIFUNGAL ACTIVITY OF PLANT-DERIVED COMPOUNDS

Standardized screening methods are essential for a successful identification of antifungal compound activity. These methods should be rapid allowing to analyze a large number of spores or to detect any growth activity of mycelium. A preliminary screening should be made under *in vitro* conditions in laboratory. This type of bioassays enables to identify any potential antifungal activity of the tested plant-derived compounds, and gives preliminary information on the efficient concentration and possible modes of action of the biofungicide. The subsequent step consists in verifying the effectiveness of the novel biofungicide under *in vivo* conditions.

Essential in all assays is the use of controls, the comparison to commercial fungicides and also the use of different concentrations of tested substances (Uldahl and Knutsen, 2009) in order to identify the lowest efficient concentration to use. Nevertheless, screening methods can be in general influenced by various factors, e.g. chemical reactivity and purity of the tested substance, time, stage of growth of the fungus and plant, change in pH, oxidation of an active compound, medium composition, nature of environment, and scale of bioassay (Paxton, 1991).

5.1. Laboratory Screening Methods

Vegetatively growing cultures exposed to biological extracts are generally used in bioassays with fungi. The minimal inhibitory concentration (MIC) of the plant-derived product is usually determined through dilution methods (Rai and Mares, 2003). *In vitro* screening methods for detecting antifungal compounds can be categorized into (1) plate dish diffusion assays on agar medium (Langvad, 1999); (2) assays of fungal growth in liquid culture, which can be measured as increase in dry weight of fungus or in optical density of the suspension caused by increased number or size of cells inside (Rai and Mares, 2003); and (3) whole tissue assays, in which the mycelial plug is placed directly on the tissue of the plant tested for antifungal properties (Ross et al., 2008).

The most common method is represented by plate dish diffusion assay. The fungal growth is estimated by measuring colonial mean radii of the fungus or inhibition zone compared to controls (Uldahl and Knutsen, 2009). One of the diffusion methods is the poisoned plate technique, which consists in the mycelial radial growth inhibition on an agar medium mixed with extracts and inoculated with an agar plug from actively growing colonies of the fungus in a Petri plate. Alternatively, the poisoned plate can be inoculated with the spore suspension to perform a spore germination test (Abou-Jawdah et al., 2002). Another diffusion method uses paper discs saturated with extracts and placed onto the agar surface in a Petri plate containing spores (Houdai et al., 2004; Mahakhant, 1998; Soltani et al., 2005). In the plate well method, spore suspensions are mixed with agar medium before plating in Petri dishes, where wells are stamped in the agar and the tested substance is added (Kellam et al., 1988). A method of stable gradient technology, marketed as ETEST, uses a plastic strip establishing a continuous gradient of the test compound in surrounding agar (Cormican and Pfaller, 1996; Serrano et al., 2003). An automatic image analysis method evaluates the viability and germination characteristics of fungal spores (Paul et al., 1993). Another method is based on spore swelling and germination in liquid culture; this last method allows screening of more extracts in a short time frame (Uldahl and Knutsen, 2009). In whole tissue assays, the mycelial plug is transferred onto the tissue pieces of the plant tested for antifungal properties placed in Petri dish (Ross et al., 2008).

However, one of the most efficient means of screening for antifungal compounds is represented by thin-layer chromatography (TLC) bioautography. This method is simple, rapid, and allows the direct localization of active constituents in plant extracts (Hostettmann and Potterat, 1997). Antimicrobial activity is estimated by absorbing chemicals or extracts onto the surface of chromatographic plates and placing them directly in contact with the medium inoculated with fungal cultures (Wedge and Nagle, 2000). The bioautography methods are divided into (1) contact bioautography, (2) immersion bioautography, and (3) direct bioautography that is considered to be the most efficient technique. In direct bioautography, a developed TLC plate containing tested substances is dipped in the suspension of microorganisms growing in the broth or this suspension is sprayed onto the plate (Horváth et al., 2004). Then, the plate is incubated and microorganisms grow directly on it. Antibacterials are applied to avoid contamination. The individuated chemical compounds with antifungal activity are often further examined for antifungal properties by microtiter plate test (Scher et al., 2004).

The antifungal activity of an extract can be determinated also by means of flow-cytometry. Green et al. (1994) have developed a flow cytometric assay for antifungal activity based on detection of increased permeability of the fungal cell membrane to propidium iodide (PI), a nucleic acid-binding fluorochrome largely excluded by intact cell membranes, after drug treatment. The flow-cytometric methods enable to identify and even separate living and dead fungal cells. The approximate spore size, granularity or shape can be also detected (Shapiro, 1995). This technique has several advantages such as short incubation time, higher precision in comparison with other screening methods, greater accuracy, and speed of analysis (Green et al., 1994).

The screening for antifungal activity can be performed also with detached leaf method. In this method, the mycelial plugs or spore suspensions are applied to the detached leaves treated with plant extracts and placed in Petri dish (Goufo et al., 2008). Then, the diameter of the lesion caused by fungus is measured.

4.2. *In Vivo* Screening Methods

In vivo screening methods follow the preliminary tests in laboratory conditions with the aim to evaluate the efficacy of the antifungal extracts also under the natural growth conditions of cultural crops. This part of determination of antifungal compounds' efficacy is a necessary step for a novel biofungicide registration as only this type of screening proves real antifungal potential of the tested product. This screening includes growth chamber, greenhouse and field trials. It is particularly important that the conditions are favourable to the fungal development. Nevertheless, the plants tested cannot be stressed by other factors (e.g., water stress), apart from the stress by fungal pathogen. In growth chamber, the environmental conditions must be well controlled and the influence of external factors minimized. However, the choice of crop species for testing is limited by the limited space in growth chamber. Therefore, in a growth chamber, the experiment on the plants of young developmental stages or seeds, are usually performed. In the greenhouse, managing of growth conditions ideal for pathogen is more difficult. However, greenhouse tests are ideal for antifungal activity screening at any growth stage. For some crops, that are cultivated exclusively in the field, the field trials are necessary to establish the real functionality of the biofungicide tested. In the field, the timing of inoculation with pathogenic fungus is fundamental as each pathogen attacks the plant especially in some plant stages, when the conditions are the most favourable.

In *in vivo* experiments, the choice of the appropriate inoculation method and the inoculum concentration are important. The most common inoculation method consists in the use of spore suspension; however, in some cases, the mycelial plugs are used for direct inoculation of wounded plant parts. Inoculum of air-borne pathogens is applied to aerial organs of the plant, while soil-borne pathogens are inoculated onto the soil surface or directly mixed with the substrate before transplanting. It is recommended to effectuate a pathogenity test prior to the antifungal screening to verify the aggressiveness of the fungal pathogen and to determinate the optimal inoculum concentration. The commercial fungicide should be used as a control for biocontrol agent activity screening, where available. The ideal period and the way of application of the biofungicide to the plants should be also determined.

In addition, it is also important to verify the potential phytotoxicity to the host plant as many of plant-derived defensive substances are toxic for the plants themselves and they are inactivated as glycosides or by polymeration or they are confined in intercellular spaces (Martínez, 2012). Therefore, the minimum efficient concentration of plant-derived product should be always determined.

5. CASE STUDY: ANTIFUNGAL ACTIVITY OF *VITEX AGNUS-CASTUS* AGAINST *PYTHIUM ULTIMUM* IN TOMATO

Pythium ultimum Trow. is one of the most common causes of damping-off and seed rots of many crops, including tomato (*Lycopersicon esculentum* Mill.), and its management is generally a difficult challenge. Antifungal activity of some plant-derived products against this fungus has been reported by some authors (Brooker et al., 2008; Mdee et al., 2009; Alhussaen et al., 2011; Yoon et al., 2011) and proposed for use in the control of the fungus.

Nevertheless, only few studies have been made on plant-derived products such as plant extracts and on their capacity to induce defence responses in tomato (Daayf et al., 2000; Medeiros et al., 2009). As suggested by Harm et al.

(2011), some natural substances, e.g. *Solidago canadensis* extract, can induce natural resistance-related metabolites such as PR-proteins to enhance tolerance of plants to pathogens. Moreover, these substances may act in synergy with fungicides, thus aiding in reducing their ecological burden.

In this case study, antifungal activity of the stem extract of *Vitex agnus-castus* L. (*Verbenaceae*) was investigated both under *in vitro* and *in vivo* conditions. *In vivo* experiments were carried out on tomato seedlings to compare the antifungal activity of this plant extract with that of the synthetic fungicide. Activation of *PR* genes after treatment with *V. agnus-castus* extract and *P. ultimum* inoculation in tomato was investigated to elucidate the mode of action of the plant extract.

Crude methanol extract from stems of *V. agnus-castus* was obtained by the Soxhlet extraction technique (Harborne, 1998). Solvent was evaporated under reduced pressure, as described by Haouala et al. (2008). For *in vivo* experiments, tomato cv. Superprecoce di Marmande (SGARAVATTI) was used. *P. ultimum* (MUCL 30159, from BCCM™, Belgium) was isolated from infected tomato plantlets.

Toxicity tests of plant extract against the fungus were performed using the poisoned plate technique, according to the procedure proposed by Abou-Jawdah et al. (2002). Plant extract sterilized by filtration through 0.2 µl filter was added to PDA after autoclaving, when the temperature of the medium reached 50°C, and mixed thoroughly. For *in vitro* experiments, the final volume of the extract in 20 ml of PDA per each Petri dish was adjusted to four different final concentrations (0.05, 0.1, 0.2, 0.4 %). PDA plates unamended with plant extracts served as controls. Mycelial growth inhibition tests were performed by placing in the centre of each plate 3 pieces of 5 mm mycelial agar discs cut from the margin of actively growing fungal colonies. Colony diameter was measured after incubation at 26°C for 3 days, and mycelial development was observed for 10 days. All treatments were replicated 5 times. The percentage of inhibition was calculated comparing treated plates with the control. The Minimum Inhibitory Concentration (MIC) was established as the lowest extract concentration that resulted in no visible mycelial growth after 20 hours. Fungicide activity was calculated when no fungal growth was observed in the plates, whereas fungistatic activity was calculated when fungal growth was delayed (Osorio et al., 2010).

In the growth chamber experiment on tomato germinating seeds, the inoculum ($4 \cdot 10^5$ CFU·ml^{-1}) was applied to the substrate, one day before sowing, at the dose of 1 ml per seed. The seedling containers were covered with plastic to maintain a high relative humidity. Treatments with *V. agnus-*

castus extract at three different concentrations (0.2, 0.4, 0.8 %; 2.5 ml per seed) were carried out one day after inoculation. Inoculated controls treated with a chemical fungicide (Previcur Energy - BAYER, 3 ml·m^{-2}) or distilled water, an uninoculated control treated with extract at 0.8 %, the highest one used, to exclude phytotoxicity caused by the extract, and an uninoculated control treated with distilled water were established. Each treatment consisted of 30 plants and was replicated three times. The treatments were repeated 7 days after sowing. The plants were kept at the temperature of 26°C and high relative humidity. The number of germinated seeds was recorded 14 days after inoculation.

For the gene expression analyses, tomato seeds were sown in a peat substrate and maintained at 25°C in the growth chamber. Extract of *V. agnus-castus* (0.8 %) at a dose of 2.5 ml per seedling (cotyledonary stage) was applied to the substrate and plant crown. One ml of the suspension containing $4 \cdot 10^5$ CFU of *P. ultimum* per plant was applied to the substrate and to plant crown immediately after extract treatment. The inoculated control treated with distilled water, uninoculated control treated with extract, and uninoculated control treated with distilled water, were established. Twenty plant samples per treatment were collected 6, 12, 24 and 48 hours after treatment, frozen immediately by liquid nitrogen, and stored at -80°C. All materials used during the collection of plant samples for RNA preparation were treated with DEPC 0.1% (v/v) and autoclaved at 121°C for one hour. RNA extraction and RT-PCR were performed as described by Švecová et al. (2013).

The results showed that increasing concentrations of *V. agnus-castus* extract have progressively delayed the mycelial growth of *P. ultimum in vitro* (Figure 1). The MIC was 0.2 % as it is the lowest concentration with clear-cut inhibitory effect in comparison with control. The extract evidenced a fungistatic effect since, at higher concentrations, the fungus restarted to grow, after 7-10 days of *in vitro* culture. In *in vivo* antifungal assays, *V. agnus-castus* extract has showed significant antifungal activity against *P. ultimum* also on tomato germinating seeds (Figure 2). Its efficacy was comparable with that of the synthetic fungicide, as reported also by Tegegne and Pretorius (2007) for some plant extract/pathogen interaction. Moreover, no phytotoxicity by plant extract was recorded. It was previously reported that the age of plants can affect the efficiency of plant extracts (Tohamy et al., 2002). Therefore, it may be supposed that higher concentrations of the *V. agnus-castus* extract should be employed to reduce disease incidence in plants at later stages of plant development, since the fungus can cause damages at the root apex of adult plants and fruits.

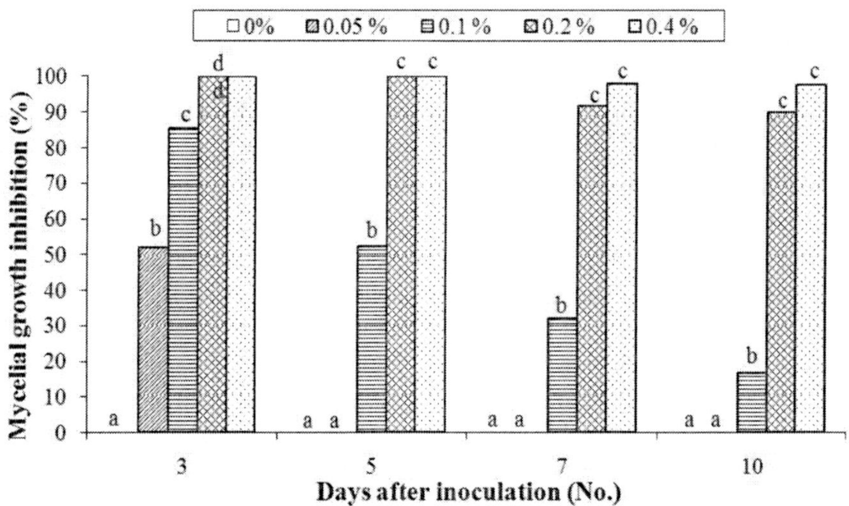

Figure 1. Inhibition of *P. ultimum* mycelial growth by 4 concentrations of *V. agnus-castus* extract over time. According to Duncan's multiple range test, different letters indicate significant differences among extract concentrations ($P < 0.05$).

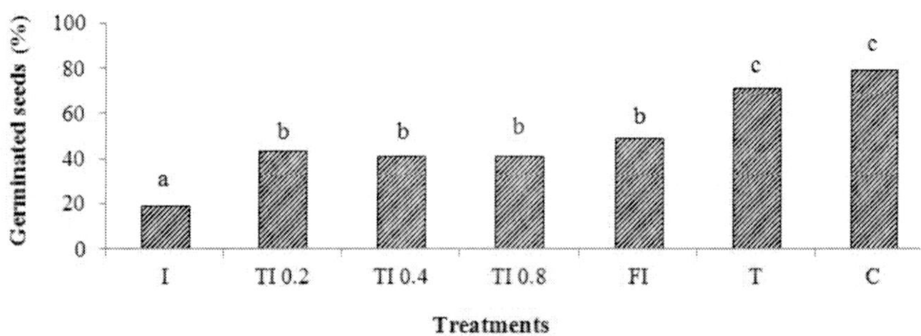

Figure 2. Germinated tomato seeds (%) after *P. ultimum* inoculation and/or *V. agnus-castus* extract treatment. According to Duncan's multiple range test, different letters indicate significant differences among treatments ($P < 0.05$). (I) seeds inoculated with *P. ultimum*; (TI 0.2) seeds treated with *V. agnus-castus* extract (0.2 %) and inoculated with *P. ultimum*; (TI 0.4) seeds treated with *V. agnus-castus* extract (0.4 %) and inoculated with *P. ultimum*; (TI 0.8) seeds treated with *V. agnus-castus* extract (0.8 %) and inoculated with *P. ultimum*; (FI) seeds treated with synthetic fungicide and inoculated with *P. ultimum;* (T) seeds treated with *V. agnus-castus* extract (0.8 %); (C) seeds treated with distilled water.

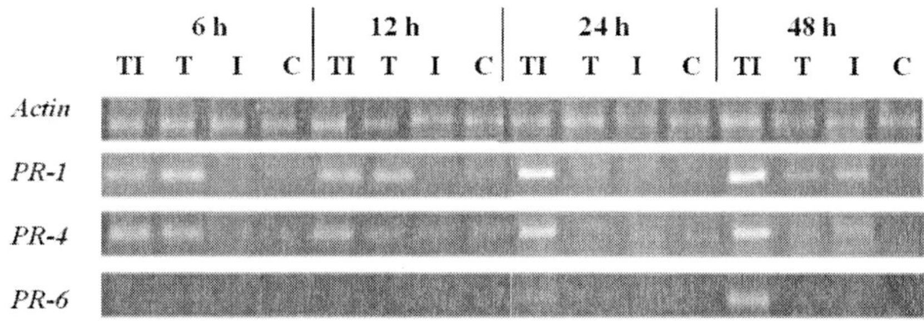

Figure 3. Time course of expression of the tomato *PR* genes in response to *P. ultimum* and/or *V. agnus-castus* extract. (TI) plants treated with *V. agnus-castus* extract and inoculated with *P. ultimum*, (T) plants treated with *V. agnus-castus* extract, (I) plants inoculated with *P. ultimum*, (C) uninoculated control plants treated with distilled water.

To determine the involvement of plant extract and pathogenic fungus in *PR* gene induction, tomato seedlings both uninoculated and inoculated with *P. ultimum* were treated with *V. agnus-castus* extract. The expression of four *PR* genes was monitored by RT-PCR at four time points within 48 hours after treatment (Figure 3). In the control plants (C) treated only with water, no significant activation of monitored *PR* genes was observed for the whole duration of the experiment.

The *PR-1* and *PR-4* genes were activated directly by *V. agnus-castus* extract up to 12 hours after treatments; at 24 hours, the direct activation by plant extract disappeared and a synergistic inducing effect of extract and pathogen applied simultaneously to the plant was observed (Švecová et al., 2013). At 24 and 48 hours, the synergistic inducing effect was observed also in the case of the *PR-6* gene, which was not activated directly at the previous time points.

Plant response regarding the activation of *PR-1* and *PR-4* genes at the 24 and 48 hours might indicate some synergistic inducing effect of the simultaneous application of plant extract and pathogen inoculum. This is confirmed also by the fact that activation of genes in the plants treated only with *V. agnus-castus* extract (T) was very weak at later time points (24 and 48 hours).

Therefore, at these time points, the *PR1* gene activation was enhanced by plant extract and pathogen, when applied together. Similar response was observed also in *PR1* gene expression of *Arabidopsis*, which was equally enhanced by growing the plants in olive compost and by *B. cinerea* inoculation, and was further boosted in compost-grown pathogen-inoculated plants (Segarra et al., 2013). In tomato, *PR1* gene was activated also by chlorogenic acid treatment (López-Gresa et al., 2011); and Perkovska et al. (2007) reported that the elicitoractive carbohydrates from cell walls of the pathogen *B. cinerea* were effective to induce systemic resistance against tomato foliar diseases.

Also other authors (Benhamou et al., 2001) suggest that applications of the fungal protein sensitize susceptible tomato plants to react more rapidly and more efficiently to *F. oxysporum* f. sp. *radicis-lycopersici* attack, mainly through the massive accumulation of fungitoxic compounds at sites of attempted pathogen penetration.

Therefore, also in our case, *P. ultimum* might have acted as an elicitor when applied simultaneously with the plant extract at a concentration of 0.8 % that was fungistatic under *in vitro* conditions (Švecová et al., 2013); in synergy with plant extract, the fungus triggered the *PR* genes activation. However, the mechanism of synergy of *V. agnus-castus extract* and *P. ultimum* has not been elucidated yet. The ability of *V. agnus-castus* extract to enhance plant defence responses upon pathogen inoculation might be further investigated. The activation of various *PR* genes suggests that induction of defence responses by *V. agnus-castus* extract in tomato may be regulated by more than one signalling pathway (Švecová et al., 2013).

According to the result of this case study, the *V. agnus-castus* extract could potentially be employed in the organic or sustainable management of *Pythium* as an alternative to conventional fungicides. The *V. agnus-castus* extract as a promising antifungal agent should be tested for *Pythium* control also in other crops to verify its efficacy in different culture systems and environments.

Moreover, expression of *PR* genes in tomato seedlings after treatment with *V. agnus-castus* extract indicates that this plant extract contributes to the activation of plant defence mechanisms. Induction of defence responses in plants by *V. agnus-castus* extract is a valuable approach for sustainable crop disease management as it is in accordance with the current trend to reduce chemical applications to crop cultures.

CONCLUSION

Recently, the interest in natural plant-derived compounds increased drastically at commercial level. Interestingly, also the companies traditionally producing synthetic fungicides are now interested in natural compounds. This trend is due to the growing organic market and consumers' request of the products with low residue content. The increased interest in these natural compounds caused recent intensive screening of natural substances.

There are many ways to produce primary source material for the production of plant-derived compounds.

The companies' interest is firstly economic. It means that the source material of low initial cost is desirable. Thus, also waste organic material is largely studied. The strategies of manufacturing the raw materials are continuously being developed. From the point of view of the costs, the production of the crude extracts is the most convenient. However, complex extraction methods for single compound separation are also being employed in order to obtain more efficient product.

Plant-derived compounds represent a great challenge for the industries. However, there are some obstacles that often do not permit the commercial exploitation of plant-derived products. It is often hard to find a stable natural product with a broad spectrum activity suitable for commercial production. Therefore, the agrochemical industry has largely focused on the design of novel, synthetic compounds based on the structure of a natural product (Martínez, 2012). These synthetic compounds ideally possess optimized biological, physical and environmental properties and are often simpler in structure than the original natural substances (Clough and Godfrey, 1998).

It is to be supposed that also in the future the research of such compounds will continue to increase in order to develop eco-friendly products for crop protection.

Therefore, the deep knowledge of the effects of these plant-derived compounds on the plant as well as on human health is essential.

REFERENCES

Abdolmaleki, M., Bahraminejad, S., Abbasi, S., 2011a. Antifungal activity of some plant crude extracts on four phytopathogenic fungi. *Journal of Medicinal Plants*, 10(38), 148-155.

Abdolmaleki, M., Bahraminejad, S., Salari, M., Abbasi, S., Panjeke, N., 2011b. Antifungal activity of peppermint (*Mentha piperita* L.) on phytopathogenic fungi. *Journal of Medicinal Plants*, 10(38), 26-34.

Abou-Jawdah, Y., Sobh, H., Salameh, A., 2002. Antimycotic activities of selected plant flora, growing wild in Lebanon, against phytopathogenic fungi. *Journal of Agricultural and Food Chemistry*, 50 (11): 3208 3213.

Agrios, G.N., 2005. Plant Pathology. Elsevier Academic Press. San Diego, USA. 922 pp.

Akinbode, O.A., Ikotun, T., 2008. Efficacy of certain plant extracts against seed-borne infection of Collectotrichum destructivum on cowpea (Vigna uniguculata). *African Journal of Biotechnology,* 7 (20): 3686-3688.

Alhussaen, K., Hussein, E.I., Al-Batayneh, K.M., Al-Khatib, M., Khateeb, W.A., Jacob, J.H., Shatnawi, M.A., Khashroum, A., Hegazy, M.I., 2011. Identification and controlling Pythium sp. infecting tomato seedlings cultivated in Jordan valley using garlic extract. *Asian Journal of Plant Pathology,* 5 (2), 84-92.

Ammar, M. I., Nenaah, G.E., Mohamed, A. H.H., 2013. Antifungal activity of prenylated flavonoids isolated from Tephrosia apollinea L. against four phytopathogenic fungi. *Crop Protection,* 49, 21-25.

Anand, T., Bhaskaran, R., 2009. Exploitation of plant products and bioagents for ecofriendly management of Chilli fruit rot disease. *Journal of Plant Protection Research,* 49 (2): 195-203.

Bailey, K.L, 2010. Canadian Innovations in Microbial Biopesticides. *Can J. Plant pathol.,* 32(2):113-121.

Bautista- Baños, S., Hernandez-Polez, M., Diaz-Perez, J.C., Cano-Ochoa, C.F., 2000. Evaluation of the fungicidal properties of plant extracts to reduce Rhizopus stolonifer of "circula" fruit (Spondias purpurea L.) during storage. Postharvest Biol Technol, 20:99-106.

Bautista-Baños, S., García-Domínguez, E., Barrera-Necha, L.L., Reyes-Chilpa, R., Wilson, C.L., 2003. Seasonal evaluation of the postharvest fungicidal activity of powders and extracts of huamuchil (*Pithecellobium dulce*): action against Botrytis cinerea, Penicillium digitatum, and Rhizopus stolonifer of strawberry fruit. *Postharvest Biology and Technology,* 29: 81-92.

Benhamou, N., Bélanger, R.R., Rey, P., Tirilly, Y., 2001. Oligandrin, the elicitin-like protein produced by the mycoparasite Pythium oligandrum, induces systemic resistance to Fusarium crown and root rot in tomato plants. *Plant Physiology and Biochemistry,* 39 (7-8), 681-696.

Bernhoft, A., 2010. A brief review on bioactive compounds in plants. In: Bioactive compounds in plants – benefits and risks for man and animals. Aksel Bernhoft editor. Oslo: The Norwegian Academy of Science and Letters, p. 12.

Bhaskara Reddy, M.V., Angers, P., Gosselin, A., Arul, J., 1997. Characterization and use of essential oil from Thymus vulgaris against

Botrytis cinerea and Rhizopus stolonifer in strawberry fruits. *Phytochemistry,* 47 (8): 1515-1520.

Bowers, J.H., Locke, J.C., 2000. Effect of botanical extracts on the population density of Fusarium oxysporum in soil and control of Fusarium wilt in the greenhouse. *Plant Disease,* 84 (3):300-305.

Brooker, N., Windorski, J., Bluml, E., 2008. Halogenated coumarin derivatives as novel seed protectants. *Communications in agricultural and applied biological sciences,* 73 (2): 81-89.

Butt, T.M., Jackson, C., Magan, N., 2001. Introduction—fungal biological control agents: progress, problems and potential. In: T.M. Butt, C. Jackson and N. Magan (Eds.), Fungi as Biocontrol Agents. Progress, Problems and Potential, CABI Publishing, Wallingford, 1–8.

Cho, J.Y., Choi, G.J., Son, S.W., Jang, K.S., lim, H.K., Lee, S.O., Sung, N.D., Cho, K.Y., Kim, J.-C., 2007. Isolation and antifungal activity of lignans from Myristica fragrans against various plant pathogenic fungi. *Pest. Manag. Sci.,* 63: 935-940.

Cho, J.Y., Kim, H.J., Choi, G.J., Jang, K.S., Lim, H.K., Lim, C.H., Cho, K.Y., Kim, J.-C., 2006. Dehydro-alpha-lapachone isolated from Catalpa ovata stems: activity against plant pathogenic fungi. *Pest Management Science,* 62 (5): 414 – 418.

Clough, J.M., Godfrey, C.R.A., 1998. The strobilurin fungicides. In: Hutsun, D. & Miyamoto, J. (Eds.), Fungicidal Activity. Chemical and Biological Approaches to Plant Protection. John Wiley & Sons Ltd., West Sussex, England, 254 pp.

Cormican, M.G., Pfaller, M.A., 1996. Standardization of antifungal susceptibility testing. *Journal of Antimicrobial Chemotherapy,* 38(4): 561-578.

Curtis, H., Noll, U., Störmann, J., Slusarenko, A.J., 2004. Broad-spectrum activity of the volatile phytoanticipin allicin in extracts of garlic (Allium sativum L.) against plant pathogenic bacteria, fungi and oomycetes. *Physiological and Molecular Plant Pathology,* 65(2), 79-89.

Daayf, F., Ongena, M., Boulanger, R., El-Hadrami, I., Belanger, R.R., 2000. Induction of phenolic compounds in two cultivars of cucumber by treatment of healthy and powdery mildew-infected plants with extracts of Reynoutria sachalinensis. *J. Chem. Ecol.,* 26:1579-1593.

Demirci, F., Dolar, F.S., 2006. Effects of some plant materials on Phytophthora blight (Phytophthora capsici leon.) of pepper. *Turkish Journal of Agriculture and Forestry,* 30(4): 247-252.

Deora, G. S., Suhalka, D., Vishwakarma, G., 2010. Antifungal potential of Philonotis revoluta - A moss against certain phytopathogenic fungi. *Journal of Pure and Applied Microbiology,* 4(1), 425-428.

Devaiah, S.P., Mahadevappa, G.H., Shetty, H.S., 2009. Induction of systemic resistance in pearl millet (Pennisetum glaucum) against downy mildew (Sclerospora graminicola) by Datura metel extract. *Crop. Protection,* 28 (9): 783-791.

Dixon R.A. 2001. Natural products and plant disease resistance. *Nature,* 411: 843-847.

Doughari, J.H., Nuya,S.P., 2008. *In vitro* antifungal activity of Deterium microcarpum. *Pak. J. Med. Sci.,* 24 (1): 91-95.

Faria, T. de J., Ferreira, R.S., Yassumoto, L., Pinto de Souza, J.R., Ishikawa, N.K., de Melo BarbosaI, A., 2006. Antifungal activity of essential oil isolated from Ocimum gratissimum L. (eugenol chemotype) against phytopathogenic fungi. *Braz. arch. biol. technol.,* 49 (6): 867-871.

Fathima, S.K., Bhat, S., Girish, K., 2009. Efficacy of some essential oils against Phomopsis azadirachtae: the incitant of die-back of neem. *J. Biopesticides,* 2, 157-160.

Francis, R. and Keinath, A., 2010. Biofungicides and chemicals for managing diseases in organic vegetable production. CLEMSON Cooperative extension. Information leaflet 88.

Freitas, G.C., Kitamura, J.O.S., Lago, J.H.G., Young, M.C.M., Guimarães, E.F., Kato, M.J., 2009. Caldensinic acid, a prenylated benzoic acid from Piper caldense. *Phytochemistry Letters,* 2: 119-122.

Goufo, P., Teugwa, Mofor, C., Fontem, D.A., Ngnokam, D., 2008. High efficacy of extracts of Cameroon plants against tomato late blight disease. *Agronomy of Sustainable Development,* 28 (4): 567-573.

Green, L., Petersen, B., Steimel, L., Haeber, P., Current, W., 1994. Rapid determination of antifungal activity by flow cytometry. *Journal of Clinical Microbiology,* 32 (4): 1088-1091.

Haouala, R, Hawala, S., El-Ayeb, A., Khanfir, R., Boughanmi, N., 2008. Aqueous and organic extracts of Trigonella foenum-graecum L. inhibit the mycelia growth of fungi. *Journal of Environmental Sciences,* 20 (12): 1453-1457.

Harborne, J.B., 1998. Phytochemical Methods: a Guide to Modern Techniques of Plant Analysis, third ed. Chapman & Hall Pub., London.

Harm, A., Kassemeyer, H.-H., Seibicke, T., Regner, F., 2011. Evaluation of chemical and natural resistance inducers against downy mildew

(Plasmopara viticola) in grapevine. *American Journal of Enology and Viticulture,* 62 (2), 184-192.

Horváth Gy., Botz, L., Kocsis, B., Lemberkovics, É., Szabó, L. Gy, 2004. Antimicrobial natural products and antibiotics detected by direct bioautography using plant pathogenic bacteria. *Acta Botanica Hungarica,* 46 (1–2): 153–165.

Hossain, M.A., Ismail, Z., Rahman, A., 2008. Chemical composition and antifungal properties of essential oils and crude extracts of Orthosiphon stamineus *Benth. Industrial Crops and Products,* 27 (3): 328-334.

Hostettman, K., Lea, J.P., 1987. Biologically active natural products. Ann. Proc. Phytochem. Soc. Eur. Oxford: Clarendon Press., 33-46.

Hostettmann, K., Potterat, O., 1997. Strategy for the isolation and analysis of antifungal, molluscicidal and larvicidal agents from tropical plants. In "Phytochemicals for Pest Control", ACS Symposium Series vol. 658, (Hedin, P.A. (Ed.); American Chemical Society, Washinghton DC, 14-26.

Houdai, T., Matsuoka, S., Matsumori, N., Murata, M., 2004. Membrane-permeabilizing activities of amphidinol 3, polyene-polyhydroxy antifungal from a marine dinoflagellate. Biochimica Et Biophysica Acta - Biomembranes, 1667(1): 91-100.

Hussin, N.M., Muse, R., Ahmad, S., Ramli, J., Mahmood, M., Sulaiman, M.R., Shukor, M.Y.A., Rahman, F.A., Aziz, K.N.K., 2009. Antifungal activity of extracts and phenolic compoundsfrom Barringtonia racemoca L. (Lecythidaceae). *African Journal of Biotechnology,* 8 (12): 2835-2842.

Hwang, Y. , Matsushita, Y., Sugamoto, K., Matsui, T., 2005. Antimicrobial effect of the wood vinegar from Cryptomeria japonica sapwood on plant pathogenic microorganisms. *Journal of Microbiology and Biotechnology,* 15(5): 1106-1109.

Kellam, S.J., 1988. Results of a large scale screening programme to detect antifungal activity from marine and freshwater microalgae in laboratory culture. *J. Br. Phycol.,* 23: 45-47.

Langvad, F., 1999. A rapid and efficient method for growth measurement of filamentous fungi. *Journal of Microbiological Methods,* 37(1), 97-100.

Li, Z.F., Zou, C.S., He, Y.Q., Mo, M.H., Zhang, K.Q., 2008. Phylogenetic analysis on the bacteria producing non-volatile fungistatic substances. The Journal of Microbiology. Volume 46, Issue 3, pp 250-256.

Liu, T., Liu, F.G., Xie, H.Q., Mu, Q., 2012. Phytopathogenic fungal inhibitors from celery seeds. *Natural Product Communications,* 7(7), 889-890.

Lopes, J.L.S., Valadares, N.F., Moraes, D.I., Rosa, J.C., Araújo, H.S.S., Beltramini, L.M., 2009. Physico-chemical and antifungal properties of protease inhibitors from Acacia plumosa. *Phytochemistry,* 70: 871-879.

López-Gresa, M.P., Torres, C., Campos, L., Lisón, P., Rodrigo, I., Bellés, J.M., Conejero, V., 2011. Identification of defence metabolites in tomato plants infected by the bacterial pathogen Pseudomonas syringae. Environ. Exp. Bot.(2011), doi:10.1016/j.envexpbot.2011.06.003.

Mahakhant, A., Padungwong, P., Arunpairojana, V., Atthasampunna, P., 1998. Control of the plant pathogenic fungus Macrophomina phaseolina in mung bean by a microalgal extract. *Phycological Research,* 46(SUPPL.), 3-7.

Martínez, J. A., 2012. Natural Fungicides Obtained from Plants, Fungicides for Plant and Animal Diseases, Dr. Dharumadurai Dhanasekaran (Ed.), ISBN: 978-953-307-804-5, InTech

Mdee, L.K., Masoko, P., Eloff, J.N., 2009. The activity of extracts of seven common invasive plant species on fungal phytopathogens. *South African Journal of Botany,* 75 (2): 375-379.

Medeiros, F.C.L., Resende, M.L.V., Medeiros, F.H.V., Zhang, H.M., Paré, P.W., 2009. Denefse gene expression induced by a coffee-leaf extract formulation in tomato. Physiological and Molecular Plant Pathology, Volume 74, Issue 2, pp 175–183

Mewari, N., Kumar, P., 2008 Antimicrobial activity of extracts of Marchantia polymorpha. *Pharmaceutical Biology,* 46 (10-11): 819-822.

Mogg, C. , Petit, P. , Cappuccino, N. , Durst, T. , McKague, C. , Foster, M., Yack, J.E., Arnason, J.T., Smith, M.L., 2008. Tests of the antibiotic properties of the invasive vine Vincetoxicum rossicum against bacteria, fungi and insects. *Biochemical Systematics and Ecology,* 36 (5-6): 383-391.

Mohana, D.C., Raveesha, K.A. , Rai, K.M.L., 2008. Herbal remedies for the management of seed-borne fungal pathogens by an edible plant Decalepis hamiltonii (Wight & Arn). *Archives of Phytopathology and Plant Protection,* 41 (1): 38-49.

Mostafa, A.A., Al-Rahmah, A.N., Abdel-Megeed, A., Sholkamy, E.N., Al-Arfaj, A.A., El-shikh, M.S., 2012. Fungitoxic properties of some plant extracts against tomato phytopathogenic fungi. *Journal of Pure and Applied Microbiology,* 66(4), 1889-1898.

Ogwulumba, S.I., Ugwuoke, K.I., Iloba, C., 2008. Prophylactic effect of paw-paw leaf and bitter leaf extracts on the incidence of foliar myco-pathogens of groundnut (Arachis hypogaea L.) in Ishiagu, Nigeria. *African Journal of Biotechnology,* 7 (16): 2878-2880.

Oliva, A., Meepagala, K.M., Wedge, D.E., Harries, D., Hale, A.L., Aliotta, G., 2003. Natural fungicides from Ruta graveolens L. leaves, including a new quinolone alkaloid. *Journal of Agricultural and Food Chemistry,* 51(4): 890-896.

Osorio, E., Flores, M., Hernández, D., Ventura, J., Rodríguez, R., Ahuilar, C.N., 2010. Biological efficiency of polyphenolic extracts from pecan nuts shell (Carya Illinoensis), pomegranate husk (Punica granatum) and creosote bush leaves (Larrea tridentata Cov.) against plant pathogenic fungi. *Industrial Crops and Products,* 31 (1): 153-157.

Pal, K.K., B. McSpadden Gardener, 2006. Biological Control of Plant Pathogens. The Plant Health Instructor. DOI: 10.1094/PHI-A-2006-1117-02.

Park, H.J., Lee, J.Y., Moon, S.S., Hwang, B.K., 2003. Isolation and anti-oomycete activity of nyasol from Anemarrhena asphodeloides rhizomes. *Phytochemistry,* 64(5): 997-1001.

Park, I.-K., Kim, J., Lee, Y.-S., Shin, S.-C., 2008. *In vivo* fungicidal activity of medicinal plant extracts against six phytopathogenic fungi. *International Journal of Pest Management,* 54 (1): 63-68.

Pârvu, M., Pârvu, A.E., Roşca-Casian, O., Vlase, L., Groza, G., 2010. Antifungal activity of Allium obliquum. *Journal of Medicinal Plants Research,* 4(2), 138-141.

Paul, G. C., Kent, C. A., Thomas, C. R., 1993. Viability testing and characterization of germination of fungal spores by automatic image analysis. *Biotechnology and Bioengineering,* 42(1): 11-23.

Paxton, J.D., 1991. Assay for antifungal activity. In: Hostemann, K. (ed.), Dey, P.M., Harborne, J.B. (series eds.) Methods in Plant Biochemistry 6. Academic Press, Harcourt Brace Jovanovich Publishers, London, p.33.

Cecilia Peluola, 2011. Biodiversity, Ecology and Toxic Principles in Plants – Case Study: Fungal Biocontrol, Biological Diversity and Sustainable Resources Use, PhD. Oscar Grillo (Ed.), ISBN: 978-953-307-706-2, InTech, Available from: http://www.intechopen.com/books/biological-diversity-and-sustainable-resourcesuse/biodiversity-ecology-and-toxic-principles-in-plants-case-study-fungal-biocontrol.

Pepeljnjak S., I. Kosalec, Z. Kalodera, D. Kustrak, 2003. Natural antimycotics from Croatian plants. In: Rai, M., D. Mares (eds.), Plant derived antimycotics. Current trends and future prospects. Hartworth Press. N. Y. 49-79 p.

Perkovska, G.Y., Sergienko, V.G., Grodzinsky, D.M., Dmitriev, A.P., 2007. Disease resistance in tomato induced by an elicitor derived from cell walls of the fungal pathogen, Botrytis cinerea. *Acta Phytopathologica Et Entomologica Hungarica,* 42(2): 245-251.

Phay, N., Higashiyama, T., Tsuji, M., Matsuura, H., Fukushi, Y., Yokota, A., Tomita, F., 1999. An antifungal compound from roots of Welsh onion. *Phytochemistry,* 52:271-274.

Plodpai, P., Chuenchitt, S., Petcharat, V., Chakthong, S., Voravuthikunchai, S.P., 2013b. Anti-Rhizoctonia solani activity by Desmos chinensis extracts and its mechanism of action. *Crop Protection,* 43, 65-71.

Plodpai, P., Petcharat, V., Chuenchit, S., Chakthong, S., Joycharat, N., Voravuthikunchai, S.P., 2013a. Desmos chinensis: A new candidate as natural antifungicide to control rice diseases. *Industrial Crops and Products,* 42(1), 324-331.

Prusky, D., 1997. Mechanisms of resistance of fruits and vegetables to postharvest diseases. In: Disease Resistance in Fruit. Proceedings of an International Workshop, Chiang Mai, Thailand: 19-33.

Rai, M., Mares, D., 2003. Plant Derived Antimycotics. Current Trends and Future Prospects. Haworth Press. *Binghamtom,* N.Y., USA, pp.587.

Ramakrishna Mission, 2008. Assessment of Bioactivity of Some Locally Available Medicinal and Aromatic Plants in Management of Pests and Diseases of Medicinal Plants in West Bengal. Medicinal Plants Institute, Ramakrishna Mission & Department of Food Processing Industries, Government of West Bengal, Kolkata, West Bengal, India.

Ross, C., Puglisi, M.P., Paul, V.J., 2008. Antifungal defenses of seagrasses from the indian River Lagoon, Florida. Aquatic Botany, 88:134-141.

Salah, M.A., Bedir, E., Toyang, N.J., Khan, I.A., Harries, M.D., Wedge, D.E., 2003. Antifungal clerodane diterpenes from Macaranga monandra (L) muell. et arg. (Euphorbiaceae). *Journal of Agricultural and Food Chemistry,* 51(26): 7607-7610.

Sati, S.C., Sati, N., Ahluwalia, V., Walia, S., Sati, O.P., 2013. Chemical composition and antifungal activity of Artemisia nilagirica essential oil growing in northern hilly areas of India. *Natural Product Research,* 27(1), 45-48.

Satish, S., Raghavendra, M.P., Raveesha, K.A., 2009. Antifungal potentiality of some plant extracts against Fusarium sp. *Archives of Phytopathology and Plant Protection,* 42 (7): 618-625.

Segarra, G., Santpere, G., Elena, G., Trillas, I., 2013. Enhanced *Botrytis cinerea* Resistance of Arabidopsis Plants Grown in Compost May Be

Explained by Increased Expression of Defense-Related Genes, as Revealed by Microarray Analysis. PLoS ONE 8(2): e56075. doi:10.1371 /journal.pone.0056075.

Scher, J.M., Speakman, J., Zapp, J., Becker, H., 2004. Bioactivity guided isolation of antifungal compounds from the liverwort *Bazzania trilobata* (L.) S.F. gray. *Phytochemistry,* 65(18): 2583-2588.

Serrano, M. C., Morilla, D., Valverde, A., Chávez, M., Espinel-Ingroff, A., Claro, R., 2003. Comparison of etest with modified broth microdilution method for testing susceptibility of Aspergillus spp. to voriconazole. *Journal of Clinical Microbiology,* 41(11), 5270-5272.

Shafique, S., Shafique, S., Bajwa, R., Akhtar, N., Hanif, S., 2011. Fungitoxic activity of aqueous and organic solvent extracts of *Tagetes erectus* on phytopathogenic fungus - Ascochyta rabiei. *Pak.J.Bot.,* 43 (1): 59-64.

Shapiro, H.M., 1995. Practical Flow Cytometry, 3rd ed. Alan R. Liss, New York.

Sharma, V., Dhancholia, S., 2008. Evaluation of fungitoxicants, plant extracts and antagonists against Botrytis elliptica causing blight of Lilium cultivars. *Annals of Biology,* 24 (1): 89-93.

Sharma, J.R., Amrate, P.K., Gill, B.S., 2010a. Leaf soft disease of *Aloe barbadensis* and its management. *J. Res. Punjab Agric. Univ.,* 47, 22e24.

Sharma, J.R., Singh, P., Saini, S.S., Gill, B.S., 2010b. Root rot of safed musali (Chlorophytum borivillianum) and its management. *J. Res. Punjab Agric. Univ.* 47, 20-21.

Siddiqui I.A., Shaukat S.S., Khan G.H., Zaki M.J., 2002. Evaluation of Argemone mexicana for control of root-infecting fungi in tomato. *Journal of Phytopathology,* 150:321-329.

Sisti, M., De Santi, M., Fraternale, D., Ninfali, P.m Scoccianti, V., Brandi, G., 2008. Antifungal activity of Rubus ulmifolius Schott standardized *in vitro* culture. LWT 41: 946-950.

Smirnov, S., Shulaev, V., Tumer, N.E., 1997. Expression of pokeweed antiviral protein in transgenic plants induced virus resistance in grafted wild-type plants independently of salicylic acid accumulation and pathogenesis-related protein synthesis. *Plant Physiol.,* 114:1113-1121.

Soltani, N., Khavari-Nejad, R.A., Yazdi, M.T., Shokravi, S., Fernández-Valiente, E., 2005. Screening of soil cyanobacteria for antifungal and antibacterial activity. *Pharmaceutical Biology,* 43(5), 455-459.

Sridhar, S.R., Rajagopal, R.V., Rajavel, R., Masilamani, S, Narasimhan, S., 2003. Antifungal activity of some essential oils. *Journal of Agricultural and Food Chemistry,* 51(26): 7596-7599.

Srivastava, K.K., Gupta, P.K., Tripathi, Y.C., Sarvate, R., 1997a. Antifungal activity of plant products on spermoplane fungi of Azadirachta indica (Neem) seeds. Indian For. 23, 157-161.

Srivastava, M., Paul, A.V.N., Rengasamy, S., Kumar, J., Parmar, B.S., 1997b. Effect of neem (Azadirachta indica A. Juss.) seed kernel extracts on the larval parasitoid, Bracon brevicornis Wes. (Hym.: Braconidae). *J. Appl. Entomol.,* 121, 51-58.

Suresh, G., Narasimhan, N. S., Masilamani, S., Partho, P. D., Gopalakrishnan, G., 1997. Antifungal fractions and compounds from uncrushed green leaves of Azadirachta indica. *Phytoparasitica,* 25(1): 33-39.

Švecová, E., Crinò, P., Colla, G., 2010. Control of Botrytis cinerea and Ascochyta rabiei by plant extracts in vegetables and chickpea. [Controllo di Botrytis cinerea e Ascochyta rabiei con estratti vegetali in specie ortive e cece]. *Italus Hortus,* 17 (2), 2010: 65-66.

Švecová, E., Proietti, S., Caruso, C., Colla, G., Crinò, P., 2013. Antifungal activity of Vitex agnus-castus extract against Pythium ultimum in tomato. *Crop Protection,* 43, 223-230.

Tabanca, N., Demirci, B., Baser, K. H. C., Aytac, Z., Ekici, M., Khan, S. I., 2006. Chemical composition and antifungal activity of Salvia macrochlamys and Salvia recognita essential oils. *Journal of Agricultural and Food Chemistry,* 54(18), 6593-6597.

Tabanca, N., Demirci, B., Crockett, S. L., Başer, K. H. C., Wedge, D. E., 2007. Chemical composition and antifungal activity of Arnica longifolia, Aster hesperius, and Chrysothamnus nauseosus essential oils. *Journal of Agricultural and Food Chemistry,* 55(21): 8430-8435.

Taiga, A., M.N. Suleiman, W. Sule, D.B. Olufolaji, 2008. Comparative in vitro inhibitory effects of cold extracts of some fungicidal plants on Fusarium oxysporium mycelium. *African Journal Of Biotechnology,* 7(18): 3306-3308.

Taiga, A., Olufolaji, D.B., 2008. *In vitro* screening of selected plant extracts for fungicidal properties against dry rot fungal of yam tubers (Dioscorea rotundata) in Kogi State, *Nigeria Research Journal of Biotechnology,* 3 (SPEC. ISS.):309-311.

Tegegne, G., Pretorius, J.C, 2007. *In vitro* and *in vivo* antifungal activity of crude extracts and powdered dry material from Ethiopian wild plants against economically important plant pathogens. *Bio. Control,* 52: 877-888.

Thomas, C., 2009. Managing Plant Diseases with Biofungicides. VirginiaTech. [cited 25-07-2013] Available at: http://pubs.ext.vt.edu /2906/2906-1298/2906-1298_pdf.pdf.

Tohamy, M.R.A., Aly, A.Z., Abd-El-Moity, T.H., Atia, M.M., Abed-Rl-Moneim, M.L., 2002. Evaluation of some plant extracts in control damping-off and mildew diseases of cucumber. *Egypt J. Phytopathol.,* 30 (2): 71-80.

Uldahl, S.A., Knutsen, G., 2009. Spore swelling and germination as a bioassay for the rapid screening of crude biological extracts for antifungal activity. *Journal of Microbiological Methods,* 79 (1): 82-88.

Vidhyasekaran, P., 2004. Concise Encyclopedia of Plant Pathology, Haworth Press, Binghamton, NY, pp. 619.

Vogt, V., Cifuente, D., Tonn, C., Sabini, L., Rosas, S., 2013. Antifungal activity in vitro and in vivo of extracts and lignans isolated from Larrea divaricata cav. against phytopathogenic fungus. *Industrial Crops and Products,* 42(1), 583-586.

Wedge. D.E., Nagle, D.G., 2000. A new 2D-TLC bioautography method for the discovery of novel antifungal agents to control plant pathogens. *J. Nat. Prod.,* 63:1050-1054.

Widmer, T.L., Laurent, N., 2006. Plant extracts containing caffeic acid and rosmarinic acid inhibit zoospore germination of Phytophthora spp. pathogenic to Theobroma cacao. *European Journal of Plant Pathology,* 115(4): 377-388.

Wilson, C.L., Solar, J.M., El Ghaouth, A., Wisniewski, M.E., 1997. Rapid evaluation of plant extracts and essential oils for antifungal activity against Botrytis cinerea. *Plant Disease,* 81(2): 204-210.

Wojdyla, A.T., 2000. Influence of some compounds on development of Sphaerotheca pannosa var. rosae. *J. Plant Protect. Res.,* 40:106-121.

Wokoma, E.C., Nwaejike, C., 2008. Antifungal effects of solvent extracts of nccm leaves on seed pathogens of groundnut. Asian Journal of Microbiology, *Biotechnology and Environmental Sciences,* 10 (4):725-731.

Wurms, K., Labbe, C., Benhamou, N., Belanger, R.R., 1999. Effects of Milsana and benzothiadiazole on the ultrastructure of powdery mildew haustoria on cucumber. *Phytopathology,* 89:728-736.

Yoon, M.-Y., Choi, G.J., Choi, Y.H., Jang, K.S., Cha, B., Kim, J.-C., 2011. Antifungal activity of polyacetylenes isolated from Cirsium japonicum roots against variol phytopathogenic fungi. *Industrial Crops and Products,* 34 (1), 882 887.

Zaccardelli, M., Campanile, F., Cammareri, M., Grandillo, S., 2011. Agronomical use of α-tomatine and crude extracts of Solanum spp. to control phytopathogenic fungi. Acta Horticulturae, Volume 914, 25 November 2011, Pages 401-404.

Zamora-Natera, F., García-López, P., Ruiz-López, M., Salcedo-Pérez, E., 2008. Composition of alkaloids in seeds of Lupinus mexicanus (Fabaceae) and antifungal and allelopathic evaluation of the alkaloid extract [Composición de alcaloides en semillas de Lupinus mexicanus (Fabaceae) y evaluación antifúngica y alelopática del extracto alcaloideo]. *Agrociencia,* 42 (2): 185-192.

Zida, D., Tigabu, M., Sawadogo, L., Odén, P.C., 2008. Initial seedling morphological characteristics and field performance of two Sudanian savanna species in relation to nursery production period and watering regimes. *Forest Ecology and Management,* 255 (7): 2151-2162.

In: Fungicides
Editors: M.N. Wheeler, B.R. Johnston

ISBN: 978-1-62948-043-5
© 2013 Nova Science Publishers, Inc.

Chapter 2

TRICYCLAZOLE AND AZOXYSTROBIN IN RICE BLAST MANAGEMENT: A REVIEW OF THEIR ACTIVITY AND PATHOGEN RESPONSES

Andrea Kunova and Paolo Cortesi[*]
Università degli Studi di Milano, Department of Food,
Environmental and Nutritional Sciences, Milano, Italy

ABSTRACT

Magnaporthe oryzae, the causal agent of the rice blast disease, is one of the most important pathogens of cultivated rice worldwide. It can cause serious epidemics resulting in substantial yield and economic losses. Rice blast management relies on the use of fungicides in many regions. At present, there are approximately thirty fungicides registered worldwide, most of which are being used in Asia. In Europe, azoxystrobin and tricyclazole were registered for application on rice at the beginning of 1990's and in Italy are being extensively used since then. In this chapter, the main chemical classes of fungicides used in rice blast management will be characterized, with emphasis on azoxystrobin and tricyclazole, which will be described in more detail.

[*] Corresponding author: Paolo Cortesi, Università degli Studi di Milano, Department of Food, Environmental and Nutritional Sciences, Via Celoria 2, 20133 Milano, Italy. E-mail: paolo. cortesi@unimi.it.

The molecular basis of their modes of action and their activity on various stages of pathogen life cycle will be defined. Despite having an important role in rice disease management, only scarce information about responses of the *M. oryzae* isolated from rice to azoxystrobin and tricyclazole are available, therefore the baseline response of the pathogen, risks and possible emergence of resistance to both fungicides will be discussed.

INTRODUCTION

Wheat, rice and maize are the three most important cultivated staple crops worldwide, with rice being the major cereal crop used primarily for human consumption, while maize and wheat are partially used also as animal feed [24]. Rice provides nutrition for approximately 50% of human population and, while traditionally regarded typical Asian food, it is a very important nutrition source also in other continents, such as Africa or Latin America.

Asian or common rice (*Oryza sativa*, L.) is one of the two cultivated rice species within the genus *Oryza*, which comprises together 23 different species [44, 134]. *O. sativa* originated most probably in the South-East Asia, while the other cultivated species *O. glaberrima* emerged in the Central Niger Delta region in Africa [44]. Rice is one of the oldest cultivated crops, and it had been domesticated according to different studies 6 600 - 13 500 years ago [40, 62, 79]. The first record of its cultivation date back around 2 800 B.C. in China [44, 99]. From its place of origin, it has spread to Sri Lanka and India where the evidence of its cultivation date back to 1 000 B.C. Only centuries later, rice was brought to Greece and adjacent Mediterranean areas during military expeditions of Alexander the Great to Asia [46]. Subsequently, from Greece and Sicily it diffused to other regions of Italy, Spain and Portugal. In the 16th century, the European colonizers introduced rice cultivation also to Latin America and Caribbean from where it found its way to the rest of the American continent. Finally, since the beginning of the 20th century, rice has been successfully cultivated also in Australia.

Nowadays, rice is being grown in most semi-tropical and tropical regions of the world. Almost 90% of rice is cultivated in Asia, followed by Africa and South America [33], however most of their production is destined for the local consumption. Out of the top 10 rice producing countries in 2011, only Brazil was not located in Asia. On the other hand, Northern America and Europe belong to the highest exporters of rice immediately after Asia, with almost 35% and 50% of their production exported in 2010 [33].

In the past fifty years, the worldwide production of rice has more than tripled, and its global harvested area has increased approximately 40% [33]. The rice yield exceeded 700 million tons in 2011 and it is presumed that the demand for its cultivation will continue to grow in the following years especially due to increasing population growth. The introduction of new high-yielding rice varieties combined with good agronomic practices, such as controlled irrigation and use of fertilizers, have contributed greatly to increasing the rice production [29, 98].

In the near future, further growth in rice productivity will depend primarily on increasing the yield per hectare as the available area under rice cultivation has been almost exhausted [101, 127]. Therefore, among various strategies to maximize rice yield, special attention needs to be directed towards prevention of potential crop losses.

Currently, a large percentage of potential rice yield gets lost annually as a consequence of diverse pest and weed infestations and diseases. Estimates of yield losses vary from year to year among different rice-growing regions, however, numerous reports calculate that on average 40% of rice production gets lost due to pests and diseases [97, 101]. Under favorable conditions, pathogen epidemics can cause even more severe damage and in some cases entire crop can be destroyed [6, 114]. Although vast numbers of rice pests and diseases have been described, only a few are of worldwide importance because they considerably reduce crop yield. Among these, rice stem borers and plant hoppers are the most serious pests, while bacterial blight and rice blast represent the two major rice diseases. For the purposes of this chapter, rice blast and its causal agent, *Magnaporthe oryzae*, will be discussed.

RICE BLAST

Rice blast, as stated above, is one of the most important plant diseases that represents a threat to global food security. Rightly it was ranked first place in a recent survey of "Top 10" fungal plant pathogens [27]. Blast is widespread among rice-growing regions of the world and until now it has been reported in more than 85 countries [43, 85]. Typically, it causes 10-30% yield loss, but more severe epidemics have been observed, devastating entire fields [6, 27, 104]. All developmental stages of rice plant can be prone to infection. Rice seedlings or young plants can be killed, or various parts of mature plants can be infected causing leaf or collar blast, and, more economically important, neck or panicle blast.

The Pathogen

The causal agent of rice blast is *Magnaporthe oryzae* B. Couch sp. nov. (anamorph *Pyricularia oryzae* Cavara), a haploid filamentous fungus belonging to ascomycetes. It was recently reclassified as a new species separated from *M. grisea* [21, 112, 122]. The new *M. oryzae* species is actually part of a species complex causing disease in a variety of *Poaceae* family members, including crops such as wheat, barley and millet [27, 47, 96, 119]. Although individual strains or races have only limited range of hosts, isolates of *M. oryzae* from rice were shown to be able to infect different barley cultivars in laboratory conditions, which further increases the importance of this plant pathogen, even though there is no clear evidence of such cross-infection in field [14, 47, 96].

Infection Process

M. oryzae is a hemibiotrophic fungal pathogen, that maintains both biotrophic and necrotrophic behavior [52, 75]. The infection initiates by attachment and germination of the three-celled conidium, the asexual spore of the fungus, on the host surface. Within few hours, the germinating spore differentiates an appressorium at the tip of the germination tube, a highly specialized dome-shaped structure, crucial for infection [11, 30]. Appressorium formation requires presence of a hard, hydrophobic surface, plant cutin monomers and absence of nutrients [41, 45, 69]. During the appressorium maturation melanin layer is deposited in its cell wall, which can support generation of elevated turgor pressure (>8MPa) due to accumulation of compatible solutes, particularly glycerol [25, 41]. Melanin is absent only from the zone in contact with the plant surface, the appressorium pore, from which a penetration peg invades the host cuticle and cell wall mainly by mechanical force, while the plasma membrane remains intact [45]. Once inside the cell, the penetration peg differentiates into invasive hyphae surrounded by the host derived extra-invasive hyphal membrane [52]. Rice cells remain alive during this stage of growth of the fungus, and previously infected cells are killed only after the pathogen has spread into the neighboring cells presumably through plasmodesmata [30, 52, 75]. Necrotic lesions appear typically 3-7 days after infection depending on environmental conditions, accompanied by the development of aerial conidiophores [112, 122].

Numerous conidia are produced under high relative humidity conditions, which serve as secondary inoculum for further spreading of the disease [54, 112, 122].

Disease Cycle and Epidemiology

Rice blast is a typical polycyclic disease that can accomplish several infection cycles during the rice growing season [20, 104]. Almost exclusively asexual reproduction of this pathogen has been observed in majority of the rice-growing regions [129], therefore it is believed that the sexual reproduction does not play a role in disease epidemiology. The sexual reproduction *in natura* has been described only in place of its origin, the Himalayan region in South-East Asia [94, 129].

In tropics, the pathogen continues to infect rice plants from one crop to another [53]. Instead, in temperate zones, *M. oryzae* overwinters in straw or other rice residues left in the field. This inoculum present in soil can start new infection cycle upon contact with rice roots, as it was shown that *M. oryzae* is able to infect rice roots and asymptomatically spread in the above-ground parts of the plant, where it can develop typical disease symptoms [75, 100]. Another important source of primary inoculum is infected rice seeds [17, 70]. In non-germinated infected seeds, *M. oryzae* completely covers the seed surface, and serves as a substrate for growth and reproduction of the fungus. In germinating seeds instead, it grows and sporulates preferentially on the embryonic end [31, 73]. In this way, blast fungus can easily get access to the emerging coleoptile and primary root of the germinating seed and subsequently colonize young seedling. Contaminated seedlings mostly die and serve as inoculum for infection of neighboring plants, which develop typical disease symptoms [31]. It was also speculated that blast can spread from its alternate hosts onto rice. However, analysis of rice and non-rice infecting populations showed that these are genetically separated and the non-rice populations most likely do not provide inoculum for rice infection in field [10].

Various factors influence rice blast development and disease epidemiology, among which are environmental conditions, susceptibility of host plants and inoculum density.

Environmental conditions, such as temperature, relative humidity and leaf-wetness, impact on all stages of pathogen life cycle. Germination of *M. oryzae* conidia can occur between 10-33 °C, with the optimum between 25-28 °C [20, 43, 109, 121].

High relative humidity (optimum between 92-96%) and a presence of free water are other factors favoring conidia germination. In water, germination may occur within 3-4 hours. The presence of free water is necessary also for subsequent steps; successful penetration and infection of the host plant does not occur without sufficient period of leaf wetness [7, 43]. Leaf wetness of 16-20 hours is necessary for successful penetration at the average temperature 15.6 °C, while 6-8 hours are sufficient at 25 °C [20]. Once inside the plant tissue, the temperature is a major factor influencing the length of the latent period and development of disease symptoms. The incubation time can be reduced from 13-18 days at 9-10 °C to 4-5 days at the optimal temperature 26-28 °C [113]. After sufficient latent period, conidiophores emerge under optimum conditions in 4-6 hours [43, 113]. Sporulation is favored by relative humidity above 89%. Maximum sporulation occurs 7-12 days after inoculation, although it can continue at low levels in the following 60 days [109]. Apart from high relative humidity, other factors such as rice genotype susceptibility, leaf age, water stress or nitrogen fertilization influence spore production [20]. Conidia are released prevalently during the night or early morning and are dispersed by wind and rain, which initiates a new disease cycle.

RICE BLAST MANAGEMENT

Rice blast is a disease of worldwide importance. Each year considerable resources are invested in its management. There has been a huge progress made in rice blast research and there are efforts to implement the acquired knowledge in the concept of integrated pest management. It combines various strategies of blast control, including the use of resistant rice varieties, cultural practices, and biological and chemical control.

Breeding of rice cultivars with effective and durable blast resistance is the most economically and environmentally feasible strategy for rice blast management [1].

At present, there are available cultivars resistant to blast, however, there is a risk that they become susceptible to blast because of its high pathogenic variability [28]. Since a long time, there have been efforts to breed new rice varieties resistant to blast, first using major resistance genes (*Pi* genes), and more recently, exploiting quantitative resistance loci in combination with broad-spectrum *Pi* genes, which could guarantee durable resistance of new rice cultivars [15, 18, 74, 90, 91].

The newest research indicates that varietal mixtures with diverse resistance genes can provide greater disease suppression [16, 42, 77, 133]. Another option to achieve durable resistance is the use of gene pyramiding, or introduction of multiple *Pi* genes in a single rice variety [60].

Even in the absence of highly resistant varieties, proper implementation of certain cultural practices can significantly reduce blast disease damage [1]. Correlation between high planting density and increased severity of blast infection has been observed [1, 20, 83]. Furthermore, nutritional factors that affect physiological status of the rice plant influence greatly rice blast severity. Excessive nitrogen use that is related with the intensification of rice production has a dramatic impact on rice blast, increasing disease incidence and severity [68, 71], while silicon has been shown to cause cell wall fortification, that negatively correlates with pathogen cell wall penetration, and results in lower blast damage [55].

Although the adoption of appropriate cultural practices, such as low plant density, water management and minimal nitrogen fertilization can reduce rice blast severity, these can have negative impact on rice productivity. Therefore, use of fungicides remains the main strategy to manage the disease, especially in areas where susceptible rice varieties are intensively grown. Each year, immense resources are spent for chemical protection of crops and control of rice blast alone accounts for more than 8% of total fungicide use, representing the world's largest fungicide market [64]. At present, at least thirty fungicides, belonging to various chemical classes, are registered for rice blast management worldwide [34, 64]. They can be applied as seed treatment to prevent infection of seedlings after germination or as foliar treatment at tillering and/ or booting stage to reduce leaf and panicle blast [1, 88]. The proper timing of application and combination or rotation of fungicides belonging to different chemical classes are key factors to ensure their efficacy and longevity, delaying the emergence of fungicide resistance in the field.

FUNGICIDES IN RICE BLAST MANAGEMENT

Although rice blast disease has been described centuries ago, no efficient chemical control methods were available until the beginning of 1900s. At this time, inorganic copper-based fungicides were extensively used for rice blast control, but were later discontinued because of their phytotoxicity and low efficacy [130]. They were replaced by organomercurial compounds, but also these have been shown to be highly toxic to environment [39, 86].

The real advance in plant protection began with the development of synthetic fungicides in the first half of the 20th century, and discovery of natural products with antifungal properties [59]. Following this section, the most important classes of fungicides used in rice blast management will be described (summarized in Table 1).

Table 1. Main fungicide groups used for rice blast management, classified according to Fungicide Resistance Action Committee (FRAC) Fungicide code list [35]

FRAC Code	FRAC Target Site Mode of Action	Chemical Group	Example Compounds
1	B1: β-tubuline assembly	Methyl Benzimidazole Carbamates (MBC)	Benomyl
			Carbendazim
3	G1: Sterol biosynthesis - Demethylation	DeMethylation Inhibitors (DMI) Triazoles	Epoxiconazole
			Hexaconazole
			Propiconazole
			Tebuconazole
6	F2: Phospholipid biosynthesis	Phosphorothiolates	Edifenphos
			Iprobenphos
		Dithiolane	Isoprothiolane
11	C3: Respiration - Cytochrome bc1, Qo site	Quinone outside Inhibitors (QoI)	Azoxystrobin
			Kresoxim-methyl
			Metominostrobin
			Oryzastrobin
			Trifloxystrobin
16.1	I1: Melanin synthesis - Reductase	Melanin Biosynthesis Inhibitors - Reductase (MBI-R)	Pyroquilon
			Tricyclazole
16.2	I2: Melanin synthesis - Dehydratase	Melanin Biosynthesis Inhibitors - Dehydratase (MBI-D)	Carpropamid
23	D2: Protein synthesis	enopyranuronic acid antibiotic	Blasticidin S
24	D3: Protein synthesis	hexopyranosyl antibiotic	Kasugamycin
M	M1: Multisite contact activity	Inorganic	Copper
P	P2: Host Plant Defense Induction	Benzisothiazole	Probenazole

Antibiotics

The use of antibiotics for plant protection has been studied mainly in Japan [108]. The first antibiotic commercially used for the control of *M. oryzae* was blasticidin S, isolated from *Streptomyces griseochromogenes* in 1955, followed by kasugamycin from *S. kasugaensis*, introduced in 1965 [78, 86]. They were shown to inhibit spore germination and mycelial growth of the pathogen, and in particular they interfere with protein synthesis. The emergence of resistant strains to both compounds was observed in the field [26]. Nowadays, the introduction of new antibiotics in agriculture is extremely limited, because of possible emergence of cross-resistance with antibiotics used in clinical practice. For this reason, the use of antibiotics in plant protection was prohibited in most states of the European Union and is very limited also in the US.

Methyl Benzimidazole Carbamates (MBC)

Benzimidazoles or MBCs, launched in 1970s, were the first broad-spectrum systemic fungicides used mainly for foliar diseases or for seed dressing [58]. Their considerable advantages to the previously used fungicides were their effectiveness at low doses, systemic action and low acute toxicity to mammals and human [81]. However, later they were found to have negative effects on human and animal reproduction and development [80]. Benzimidazoles inhibit fungal microtubule assembly by binding to the β-tubulin subunit [23]. As a consequence they disrupt nuclear division, germ tube elongation and mycelium growth. Because of their single-site mode of action field resistance emerged soon after their introduction in practice, especially for polycyclic pathogens [105]. Compounds such as benomyl, carbendazim or thiophanate methyl are applied as seed dressing, soil drench or foliar sprays predominantly in Asia, while they are not in use anymore in Europe [64].

Triazoles

Triazoles, or demethylation inhibitors (DMI), belong to the most important and largest group of synthetic fungicides [81]. They are part of the larger class of sterol biosynthesis inhibitors (SBI).

Triazoles are potent inhibitors of 14α-demethylase (CYP51) involved in biosynthesis of ergosterol, an important component of fungal cell membrane [61, 126]. Their application results in abnormal fungal growth and eventually death of the pathogen [82]. DMI growth-regulatory effects on plants have been also observed, such as inhibition of gibberellin and plant sterol synthesis, resulting in stunted growth [12]. The pathogen resistance to triazoles is based on mutations in multiple genes and is characterized by slow shifting in pathogen sensitivity, resulting in intermediate levels of resistance and incomplete cross-resistance with other DMI fungicides [58]. DMIs used for rice blast management are flutriafol, propiconazole, epoxiconazole, hexaconazole, tebuconazole, etc. [19, 32].

Phosphorothiolates

Organophosphorous compounds represent a historical class of pesticides introduced on the market in 1960s [58]. Although most of them are insecticides, phosphorothiolates were shown to be effective fungicides against rice diseases, mainly blast [95, 118]. Their proposed mode of action is disruption of choline biosynthesis, a component of a major membrane phospholipid, phosphatidyl choline [76, 118]. They are therefore classified also as choline biosynthesis inhibitors (CBI). CBI fungicides iprobenfos, edifenphos and isoprothiolane are mainly used for rice blast management as soil drench or foliar sprays especially in Asian countries such as Japan, Korea, India or China. As for other fungicide classes, field resistance to CBI fungicides has been observed. In laboratory conditions, *M. oryzae* resistant strains were selected, which simultaneously showed increased sensitivity to SBIs [107]. However, this negative cross-resistance has not been confirmed for field isolates [57].

Quinone Outside Inhibitors (QoI)

The discovery of QoIs meant a new revolution in fungicide development, as they are derived from a natural product strobilurin A, isolated from the fungus *Strobilurus tenacellus* in 1977 [3]. At present they are the most significant chemical group introduced to the pesticide market since DMIs [93]. The first commercial products were launched in early 1990s by BASF (kresoxim-methyl) and Zeneca/Syngenta (azoxystrobin) [93].

Strobilurins have an unusually broad spectrum comprising diverse Ascomycetes, Deuteromycetes and Oomycetes [58]. They inhibit mitochondrial respiration by blocking electron transfer at the Quinone outside (Qo) site of cytochrome *bc1* complex in the inner mitochondrial membrane [8]. Because of their site-specific mode of action, several pathogens resistant to QoI fungicides have been detected; the first only two years after their commercialization [8].

Out of approximately 20 QoI compounds available, azoxystrobin is the most widely used for rice blast management, including EU, US, Latin America, Japan, India and other rice growing regions. The other strobilurin-based fungicides include oryzastrobin, motominostrobin, trifloxystrobin etc.

Melanin Biosynthesis Inhibitors (MBI)

MBI fungicides are particular fungicides; they do not interfere with functions related to fungal viability, but they act on plant-pathogen interaction [58]. They are almost exclusively used for rice blast management. They specifically interfere with melanin accumulation in the appressorium which impedes subsequent host cell-wall penetration.

Two different sub-classes of MBI have been characterized, differing in the main enzymatic process they target; dehydration (MBI-D) or reduction (MBI-R) steps of melanin biosynthesis [9, 67]. MBI-R compounds tricyclazole, pyroquilon and phthalide were introduced to the pesticide market in 1970-80s [58, 81]. They bind 1,3,8-trihydroxynapthalene reductase (3HNR), which blocks the conversion of trihydroxynapthalene (3HN) to vermelone [115]. On the other hand, MBI-D subgroup includes compounds carpropamid, dichlocymet and fenoxanil, which specifically inhibit scytalone dehydratase (SDH) catalyzing dehydration of scytalone into 3HN [67].

Because MBIs do not interfere with viability of *M. oryzae*, low risk of resistance development was expected for these fungicides. On the contrary, shortly after their introduction in use, MBI-D resistant strains emerged due to a single point mutation in the SDH gene [50, 111, 125].

However, no resistance to MBI-R subgroup has been observed in the field [67, 104]. Only several low-level tricyclazole-resistant strains of *M. oryzae* were obtained in laboratory, all of which showed decreased fitness, i.e. reduced sporulation and pathogenicity to rice [132].

Plant Defense Activators (PDA)

In several compounds strong *in vivo* antimicrobial activity has been observed, while they showed very weak or none *in vitro* pathogen inhibition. These substances have been proposed to induce host plant defenses, most probably via the systemic acquired resistance pathway (SAR) [106]. Although this phenomenon has been widely studied, only few of the potential PDAs have been commercialized.

Probenazole formulated as Oryzamate® has been used on rice against blast predominantly in Japan since 1975 [48, 63]. Moreover, it exhibits some antibacterial and antiviral properties and has been used in several other crops against bacterial diseases [106]. Despite its extensive use in some countries, no resistance in *M. oryzae* has been observed [48].

In the first part of the chapter the major classes of fungicides used against rice blast were described. Most of these compounds are being used in Asia, where the growers have possibility to choose and alternate products with different modes of action.

On the contrary, blast fungicide market is extremely limited in Europe, relying mainly on the use of azoxystrobin and tricyclazole. The latter, currently used under temporary permission, is in the process of its re-registration.

In the following section, these two fungicides, probably the most widely used and active against rice blast, will be described in more detail. Their biochemical modes of action, and biological activities on specific stages of pathogen development will be revisited with the addition of the newest knowledge.

AZOXYSTROBIN

Azoxystrobin is one of the first synthetic strobilurins, obtained by chemical modifications of the naturally occurring, light-sensitive strobilurin A. It was originally developed in late 1980s and was launched on the pesticide market in early 1990s by Zeneca (now Syngenta) [93].

Azoxystrobin is currently one of the best selling fungicides worldwide, it is registered in over 70 countries for more than 85 different crops [110]. It targets pathogens from all major groups of fungi, including Ascomycetes, Basidiomycetes, Deuteromycetes and Oomycetes [8].

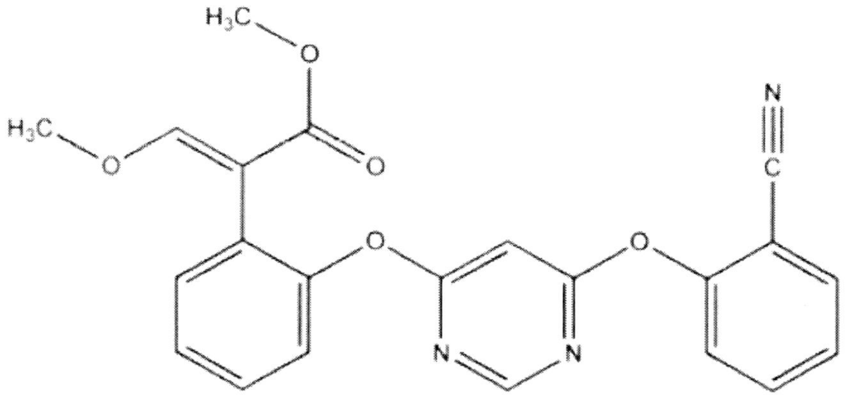

Figure 1. Molecular structure of azoxystrobin (methyl (αE)-2-[[6-(2-cyanophenoxy)-4-pyrimidinyl] oxy]-α-(methoxymethylene)benzeneacetate).

Mode of Action

Azoxystrobin, similar to other QoI fungicides, inhibits fungal mitochondrial respiration by binding to the Qo-site of the cytochrome *b*. By this interaction the electron transfer is disrupted between the cytochrome *b* and *c1* in the complex III within the inner mitochondrial membrane [8].

The inhibition of electron transfer results in loss of ATP synthesis and the disruption of the cell energy cycle.

Moreover, Qo-site is an important site for O_2 reduction to generate reactive oxygen species (ROS), which cause DNA and protein damage leading to cell death [22].

Indeed, increased production of ROS has been observed in some fungal pathogens upon azoxystrobin treatment [51]. In plants and many fungal species, disruption of mitochondrial respiration can be at least partially complemented by activation of alternative respiration [4].

The alternative oxidase (AOX), indeed, can be activated by ROS, thus protecting mitochondria and cells during the disturbance of cytochrome mediated mitochondrial respiration [128]. The existence of this alternative mechanism and its implications on biological activity and resistance development against QoIs in *M. oryzae* will be discussed subsequently.

Biological Activity

Because of its biochemical mode of action, azoxystrobin, as other QoIs, affects major energy-requiring processes in the life cycle of target fungal pathogens. Particularly, spore germination and zoospore motility are extremely sensitive to QoIs because of their high demand for energy [8]. Moreover, mycelium growth and spore formation are impaired after QoI treatment.

Despite its significance, only scarce data about the activity of azoxystrobin on diverse phases of *M. oryzae* life cycle are available, including only a limited number of isolates. Generally, similar effects to other fungal pathogens are presumed. Laboratory studies with *M. oryzae* isolates obtained from rice, perennial ryegrass, or barley demonstrate inhibition of conidial germination, mycelial growth or spore formation by azoxystrobin (Table 2). In these experiments, salicylhydroxamic acid (SHAM), the inhibitor of AOX, has been often used with varying effects.

The addition of SHAM has been justified by the observation, that although the AOX-mediated rescue mechanism is active during saprophytic stages of pathogen life, it has little impact on the field performance of QoIs, presumably due to plant antioxidants quenching ROS accumulation that would lead to AOX activation [5].

However, a study using *M. oryzae* mutant strains in AOX showed that alternative oxidation has importance *in planta* and it provides the pathogen protection from azoxystrobin treatment during the plant infection [4]. Nonetheless, the ability of AOX to rescue from azoxystrobin treatment varied with the process examined; while it was effective during mycelial growth both *in vitro* and *in planta*, it was weak during sporulation *in planta* [4, 124].

These experiments encourage careful interpretation of results when evaluating effects of azoxystrobin, and other QoIs together with SHAM. While addition of SHAM during sporulation might mimic the absence of alternative oxidation upon azoxystrobin treatment, its inclusion in mycelial growth and conidia germination might result in underestimating fungicide effective doses.

Resistance Risk

Azoxystrobin, and QoIs in general, are fungicides with high risk of resistance development [38]. The first pathogen that developed resistance to QoI in field was *Blumeria graminis* f.sp. *tritici*, in which resistant strains emerged only two years after QoI use in field [72, 103].

Table 2. Sensitivities of conidia germination, mycelium growth and sporulation of *M. oryzae* isolates from different hosts to azoxystrobin

Host plant	No. isolates	ED_{50}* (mg L^{-1} azoxystrobin)						Reference
		Conidia germination		Mycelial growth		Sporulation		
		-SHAM	+SHAM	-SHAM	+SHAM	-SHAM	+SHAM	
Rice	5	nd[†]	nd	0.044	nd	0.017	nd	[65]
Perennial ryegrass	111	nd	0.039	nd	nd	nd	nd	[72]
not specified	1	0.094	0.004	0.013	0.007	nd	nd	[49]
Goosegrass Weeping lovegrass Barley	1	nd	nd	1.8	0.01	nd	nd	[4]
Goosegrass Weeping lovegrass Barley	1	nd	nd	7.1 *in planta*	0.2 *in planta*	0.2 *in planta*	0.1 *in planta*	[4]
rice	2	>10	0.015/0.017	0.04/0.05	0.005/0.007	nd	nd	[5]
Goosegrass Weeping lovegrass Barley	2	>10	0.008/0.013	0.07/1.12	0.008/0.01	nd	nd	[5]
Rice	1	1.133	<0.001	0.147	<0.005	nd	nd	[87]
Perennial ryegrass	21	nd	0.029	nd	nd	nd	nd	[120]

ED_{50}* - the concentration of azoxystrobin causing 50% inhibition of the respective biological process.

nd[†] - not determined in the study.

At present, approximately 40 different fungal pathogens have been identified that exhibit field resistance to QoIs, including *M. oryzae* from perennial ryegrass and most recently, also *M. oryzae* from rice fields in Japan [36, 56, 120].

Three different mutations in cytochrome *b* are known to confer partial or complete QoI-resistance, substitution of glycine to alanine at position 143 (G143A), phenylalanine to leucine at position 129 (F129L) or glycine to arginine at position 137 (G137R) [37].

The substitution G143A is the most frequent and the strongest, conferring complete resistance and is associated with complete loss of QoI fungicide efficiency. Its resistance factor (RF=ED_{50}-resistant strain/ED_{50}-sensitive strain) is usually greater than several hundreds [37, 56].

On the other hand, F129L and G137R are less frequent, with the second one been identified only in *Pyrenophora tritici-repentis* with a very low frequency [102]. Their RF is rarely higher than 50, and therefore a proper application of QoI fungicides still provides efficient control of these strains. In *M. oryzae*, mutations G143A and F129L were described [36, 56].

TRICYCLAZOLE

Tricyclazole is the main representative of MBI-R fungicides, almost exclusively used against rice blast. It was developed in 1970s and is probably the most widely used fungicide in rice paddy [39, 89]. Although extensively used in the wetland derived ecosystems, it exhibits low toxicity to non-target aquatic organisms mainly because of its specific inhibition of DHN-melanin biosynthesis found mostly in fungal species [92]. It is used predominantly in Asian countries, especially in Japan, China and India; and in some countries of Latin America. It is not registered in the US and at present is undergoing re-registration in the EU.

Mode of Action

Tricyclazole's mode of action has been studied in depth in *M. oryzae*, where it has been shown to interfere with reduction steps in melanin biosynthesis. Its application mimics the *buff-* mutation, a genetic defect in 3HNR resulting in incapacity to convert 3HN into vermelone [123].

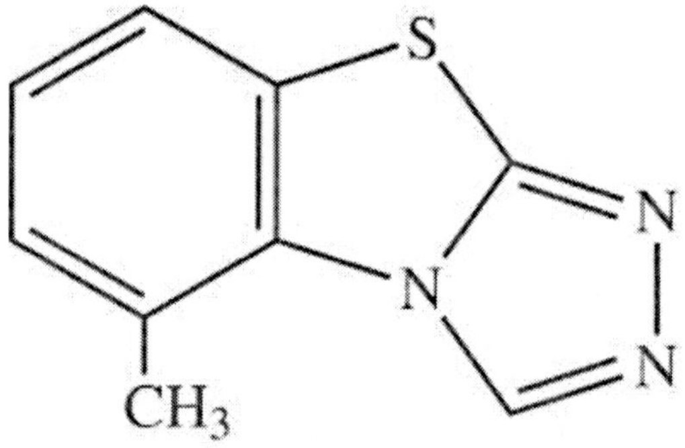

Figure 2. Molecular structure of tricyclazole (5-methyl-1,2,4-triazolo[3,4-*b*] benzothiazole).

Because of its chemical structure, tricyclazole competitively interferes with binding of 3HNR substrates [67, 115]. The direct interaction between tricyclazole and 3HNR-NADPH complex has been shown by steady-state kinetics and titration studies and its binding site within the 3HN-reductase active centre has been elucidated [2, 115]. Later, a second physiological target of tricyclazole was discovered, a 1,3,6,8-tetrahydroxynaphthalene reductase (4HNR), that has approximately 200-fold lower affinity to this fungicide [116]. At high tricyclazole concentrations, typical cherry red pigment can be detected in the growth medium of wild-type *M. oryzae* strains, due to accumulation of 4HN oxidation byproducts after both 3HNR and 4HNR become deactivated [116].

Biological Activity

Melanins can be found in many fungal species, playing roles in protecting the hyphae, spores and dormancy propagules from variety of adverse environmental conditions, such as desiccation, ultraviolet light irradiation or attack by other microbial antagonists [13]. In *Magnaporthe* and *Colletotrichum* spp., melanin biosynthesis is indispensable for pathogen penetration into the host tissues. It accumulates specifically in the cell wall of the specialized infection structure, appressorium, where it creates a barrier to solute efflux [45]. As a consequence of solute accumulation in appressorium, turgor pressure of approximately 8 MPa is generated, which leads to the mechanical penetration of the host surface [25, 41].

As tricyclazole specifically inhibits DHN-melanin synthesis, its main biological activity is in preventing penetration of *M. oryzae* into the host tissues. It has very low fungicidal activity, and it inhibits mycelium growth of the pathogen only at high concentrations *in vitro* (ED_{50} = 100.4 mgL^{-1}) [39, 65]. Similarly, spore germination, tube elongation or appressorium formation are not inhibited at low concentrations, which further confirms its low fungitoxicity [39, 65]. Tricyclazole has excellent preventative activity *in vivo*, as it blocks lesion formation at low concentrations (ED_{50} = 0.46 mgL^{-1}) [130].

Unexpectedly, tricyclazole was shown to decrease spore formation on fungicide exposed mycelia, even though no direct connection between melanin synthesis and sporulation is known at present [65]. It is possible that inhibition of melanin biosynthesis in some way interferes also with sporulation, or that apart from 3HNR and 4HNR tricyclazole targets some other enzyme involved in spore formation [65].

In field conditions, reduced secondary infection was observed after tricyclazole application [84]. This could be explained partially by inhibition of sporulation by unknown mechanism. However, reduced infection ability of spores produced from mycelia exposed to tricyclazole was described, although their germination was not compromised [65, 84]. Similar effects have been observed in *M. oryzae* after pyroquilon treatment, another fungicide belonging to MBI-R [117]. The mechanism underlying this phenomenon is still not clear.

Resistance Risk

MBI fungicides are group with non-direct fungitoxic activity, therefore low risk of resistance development was expected after their introduction to use. However, because of their specific mode of action, resistance emergence is still possible. Indeed, in carpropamid, a representative of MBI-D group targeting SDH, occurrence of resistant strains was detected only three years after its commercial use, associated with a single mutation in the SDH gene [111, 125].

On the other hand, in the historically older MBI-R group, no mutations conferring stable resistance have been described even after 30 years of their use in some regions. Decreased field performance of tricyclazole was described in some rice growing regions in China, that brought major concerns about tricyclazole resistance development. However, following analyses did not confirm either the stability of the decreased sensitivity in *M. oryzae* strains, or the presence of the mutation in the 3HNR or 4HNR genes [130, 131]. Moreover, the attempts to obtain tricyclazole-resistant strains by UV-mutagenesis in laboratory conditions were not successful [132]. The authors of the study identified only three low-level resistant strains with decreased fitness and impaired sporulation and pathogenicity. The molecular analysis of the two reductase genes did not reveal mutations in their sequence and therefore it was concluded that the presumed resistance was conferred by a mutation in different gene [132]. Similarly, a recent study analyzing possible adaptation of Italian populations of *M. oryzae* from fields repeatedly treated with tricyclazole did not reveal shift in the fungicide sensitivity, nor isolates selected from spores germinating on media with high fungicide concentration did show decreased sensitivity to tricyclazole [66].

These reports indicate that tricyclazole still confers effective protection against rice blast and suggest that there could be another unknown mechanism underlying its long time efficacy.

CONCLUSION

In this chapter we attempted to provide an extensive overview of compounds employed in rice blast management. The most significant blasticides were described, that to our knowledge are important both from historical point of view and because of their worldwide usage. In the second part, particular attention was dedicated to two important fungicides, azoxystrobin, currently the best selling broad-spectrum antifungal product on the market; and tricyclazole, the most widely used fungicide employed exclusively against rice blast. Their modes of action were described with the addition and implementation of the newest knowledge, and the implications on the biological activity and inhibition of different stages of the pathogen life cycle were explained.

Reviewed by Casiana M. Vera Cruz, International Rice Research Institute (IRRI), DAPO Box 7777, Metro Manila, Philippines, c.veracruz@irri.org

REFERENCES

[1] Ahn, S. W. 1994. International collaboration on breeding for resistance to rice blast. Pages 137-153 In: *Rice blast disease*. R. S. Zeigler, S. A. Leong and P. S. Teng, eds. CAB International.

[2] Andersson, A., et al. 1996. Crystal structure of the ternary complex of 1,3,8-trihydroxynaphthalene reductase from *Magnaporthe grisea* with NADPH and an active-site inhibitor. *Structure* 4:1161-1170.

[3] Anke, T., et al. 1977. The strobilurins-new antifungal antibiotics from the basidiomycete *Strobilurus tenacellus*. *The Journal of antibiotics* 30: 806.

[4] Avila-Adame, C. and Koller, W. 2002. Disruption of the alternative oxidase gene in *Magnaporthe grisea* and its impact on host infection. *Mol. Plant Microbe In.* 15:493-500.

[5] Avila-Adame, C. and Koller, W. 2003. Impact of alternative respiration and target-site mutations on responses of germinating conidia of *Magnaporthe grisea* to Qo-inhibiting fungicides. *Pest. Manag. Sci.* 59: 303-309.

[6] Baldacci, E. and Corbetta, G. 1960. "Brusone" in rice during the year 1959. *Fertilité* 11:23-24.

[7]   Barksdale, T. H. and Asai, G. N. 1965. Minimum dew period and temperature required for infection by *Piricularia oryzae*. *Phytopathology* 55:503.

[8]   Bartlett, D. W., et al. 2002. The strobilurin fungicides. *Pest. Manag. Sci.* 58:649-662.

[9]   Bell, A. A. and Wheeler, M. H. 1986. Biosynthesis and functions of fungal melanins. *Annu. Rev. Phytopathol.* 24:411-451.

[10]   Borromeo, E. S., et al. 1993. Genetic differentiation among isolates of *Pyricularia* infecting rice and weed hosts. *Phytopathology* 83:393-399.

[11]   Bourett, T. M. and Howard, R. J. 1990. *In vitro* development of penetration structures in the rice blast fungus *Magnaporthe grisea*. *Can. J. Bot.* 68:329-342.

[12]   Buchenauer, H. 1995. DMI-fungicides - side effects on the plant and problems of resistance. In: *Modern selective fungicides*. 2nd ed. H. Lyr, ed. Gustav Fischer Verlag, New York.

[13]   Butler, M. J., et al. 2005. Degradation of melanin or inhibition of its synthesis: are these a significant approach as a biological control of phytopathogenic fungi? *Biological Control* 32:326-336.

[14]   Bvindi, C. N. 2010. Host status and genetic analysis of blast (*Magnaporthe oryzae*) resistance in barley. *Master of Science Research Thesis*. Wageningen University and Research Centre, Wageningen.

[15]   Carrillo, M. G. C., et al. 2009. Phylogenomic Relationships of Rice Oxalate Oxidases to the Cupin Superfamily and Their Association with Disease Resistance QTL. *Rice* 2:67-79.

[16]   Castilla, N. P., et al. 2010. Assessing the effect of resistant-susceptible associations and determining thresholds for associations in suppressing leaf and neck blast of rice. *Crop Protection* 29:390-400.

[17]   Chadha, S. and Gopalakrishna, T. 2006. Detection of *Magnaporthe grisea* in infested rice seeds using polymerase chain reaction. *J. Appl. Microbiol.* 100:1147-1153.

[18]   Chen, X., et al. 2009. Genetic and molecular analyses of blast resistance in a universal blast resistant variety, Digu. Pages 149-159 In: *Advances in Genetics, Genomics and Control of Rice Blast Disease*. G.-L. Wang and B. Valent, eds. Springer Netherlands.

[19]   Chen, Y., et al. 2013. Effect of epoxiconazole on rice blast and rice grain yield in China. *Eur. J. Plant Pathol.* 135:675-682.

[20]   Cortesi, P. 2011. Il brusone del riso: Danni causati dalle epidemie e ottimizzazione della difesa con triciclazolo. Pages 87 - 140 In: *Il ruolo economico del triciclazolo nella risicoltura italiana*. AGRA, Roma.

[21] Couch, B. C. and Kohn, L. M. 2002. A multilocus gene genealogy concordant with host preference indicates segregation of a new species, *Magnaporthe oryzae*, from *M. grisea*. *Mycologia* 94:683-693.

[22] Crofts, A. R. 2004. The cytochrome bc1 complex: Function in the context of structure. *Annual Review of Physiology* 66:689-733.

[23] Davidse, L. C. 1986. Benzimidazole fungicides: Mechanism of action and biological impact. *Annu. Rev. Phytopathol.* 24:43-65.

[24] Dawe, D., et al. 2010. Emerging trends and spatial patterns of rice production. Pages 15-36 In: *Rice in the global economy: Strategic research and policy issues for food security*. S. Pandey, D. Byerlee, D. Dawe, A. Dobermann, S. Mohanty, S. Rozelle and B. Hardy, eds. IRRI (International Rice Research Institute).

[25] De Jong, J. C., et al. 1997. Glycerol generates turgor in rice blast. *Nature* 389:244-245.

[26] Deacon, J. 2005. Principles and practice of controlling fungal growth. Pages 338-355 In: *Fungal Biology*. Blackwell Publishing Ltd.

[27] Dean, R., et al. 2012. The Top 10 fungal pathogens in molecular plant pathology. *Molecular Plant Pathology* 13:414-430.

[28] Dean, R. A., et al. 2005. The genome sequence of the rice blast fungus *Magnaporthe grisea*. *Nature* 434:980-986.

[29] Dong, K., et al. 2010. A characterization of rice pests and quantification of yield losses in the japonica rice zone of Yunnan, China. *Crop Protection* 29:603-611.

[30] Ebbole, D. J. 2007. *Magnaporthe* as a model for understanding host-pathogen interactions. *Annu. Rev. Phytopathol.* 45:437-456.

[31] Faivre-Rampant, O., et al. 2012. Transmission of rice blast from seeds to adult plants in a non-systemic way. *Plant Pathol.* 62:879-887.

[32] Fang, M. Y., et al. 2009. Sensitivity of *Magnaporthe grisea* to the sterol demethylation inhibitor fungicide propiconazole. *Journal of Phytopathology* 157:568-572.

[33] FAO. 2013. FAOStat online.

[34] FRAC. 2012. 2012 FRAC Fungicide comon names. www.frac.info.

[35] FRAC. 2013. FRAC Code List 2013: Fungicides sorted by mode of action. www.frac.info.

[36] FRAC. 2012. List of pathogens with field resistance towards QoI fungicides (Status 06/12/2012). www.frac.info.

[37] FRAC. 2006. Mutations associated with QoI-resistance (Status 12/2006). www.frac.info.

[38] FRAC. 2005. Pathogen risk list. www.frac.info.

[39] Froyd, J. D., et al. 1976. Tricyclazole: A new systemic fungicide for control of *Pyricularia oryzae* on rice. *Phytopathology* 66:1135-1139.

[40] Fuller, D. Q., et al. 2009. The domestication process and domestication rate in rice: spikelet bases from the lower Yangtze. *Science* 323:1607-1610.

[41] Galhano, R. and Talbot, N. J. 2011. The biology of blast: Understanding how *Magnaporthe oryzae* invades rice plants. *Fungal Biology Reviews* 25:61-67.

[42] Garrett, K. A. and Mundt, C. C. 1999. Epidemiology in mixed host populations. *Phytopathology* 89:984-990.

[43] Greer, C. A. and Webster, R. K. 2001. Occurrence, distribution, epidemiology, cultivar reaction, and management of rice blast disease in California. *Plant Dis.* 85:1096-1102.

[44] Gustafsson, Å. and Gadd, I. 1966. Mutations and crop improvement. VII. The genus *Oryza* L. (Gramineae). *Hereditas* 55:273-357.

[45] Howard, R. J. and Valent, B. 1996. Breaking and entering: Host penetration by the fungal rice blast pathogen *Magnaporthe grisea*. *Annual Review of Microbiology* 50:491-512.

[46] Huggan, R. 1995. Co-evolution of rice and humans. *GeoJournal* 35:262-265.

[47] Inukai, T., et al. 2006. RMo1 confers blast resistance in barley and is located within the complex of resistance genes containing Mla, a powdery mildew resistance gene. *Mol. Plant Microbe In.* 19:1034-1041.

[48] Iwata, M. 2001. Probenazole - a plant defence activator. *Pesticide Outlook* 12:28-31.

[49] Jin, L.-h., et al. 2009. Activity of azoxystrobin and SHAM to four phytopathogens. *Agric. Sci. China* 8:835-842.

[50] Kaku, K., et al. 2003. Diagnosis of dehydratase inhibitors in melanin biosynthesis inhibitor (MBI-D) resistance by primer-introduced restriction enzyme analysis in scytalone dehydratase gene of *Magnaporthe grisea*. *Pest. Manag. Sci.* 59:843-846.

[51] Kaneko, I. and Ishii, H. 2009. Effect of azoxystrobin on activities of antioxidant enzymes and alternative oxidase in wheat head blight pathogens *Fusarium graminearum* and *Microdochium nivale*. *J. Gen. Plant Pathol.* 75:388-398.

[52] Kankanala, P., et al. 2007. Roles for rice membrane dynamics and plasmodesmata during biotrophic invasion by the blast fungus. *Plant Cell* 19:706-724.

[53] Kato, H. 2001. Rice blast disease. *Pesticide Outlook* 12:23-25.

[54] Kim, S., et al. 2009. Homeobox transcription factors are required for conidiation and appressorium development in the rice blast fungus *Magnaporthe oryzae. PLoS Genet.* 5:e1000757.

[55] Kim, S. G., et al. 2002. Silicon-induced cell wall fortification of rice leaves: A possible cellular mechanism of enhanced host resistance to blast. *Phytopathology* 92:1095-1103.

[56] Kim, Y. S., et al. 2003. Field resistance to strobilurin (Q(o)I) fungicides in *Pyricularia grisea* caused by mutations in the mitochondrial cytochrome b gene. *Phytopathology* 93:891-900.

[57] Kim, Y. S., et al. 2008. Variation in sensitivity of *Magnaporthe oryzae* isolates from Korea to edifenphos and iprobenfos. *Crop Protection* 27: 1464-1470.

[58] Klittich, C. J. 2008. Milestones in fungicide discovery: Chemistry that changed agriculture. In: *Online. Plant Health Progress.* doi:10.1094/ PHP-2008-0418-01-RV.

[59] Knight, S. C., et al. 1997. Rationale and perspectives on the development of fungicides. *Annu. Rev. Phytopathol.* 35:349-372.

[60] Koide, Y., et al. 2010. Development of pyramided lines with two resistance genes, Pish and Pib, for blast disease (*Magnaporthe oryzae* B. Couch) in rice (*Oryza sativa* L.). *Plant Breeding* 129:670-675.

[61] Köller, W. 1987. Isomers of sterol synthesis inhibitors: Fungicidal effects and plant growth regulator activities. *Pesticide Science* 18:129-147.

[62] Kovach, M. J., et al. 2007. New insights into the history of rice domestication. *Trends in Genetics* 23:578-587.

[63] Kuck, K.-H. and Gisi, U. 2008. FRAC mode of action, classification and resistance risk of fungicides. Pages 415-432 In: *Modern Crop Protection Compounds.* Wiley-VCH Verlag GmbH.

[64] Kumar, M. K. P., et al. 2013. Impact of fungicides on rice production in India. Pages 77-98 In: *Fungicides - showcases of integrated plant disease management from around the world.* M. Nita, ed. InTech.

[65] Kunova, A., et al. 2013. Impact of tricyclazole and azoxystrobin on growth, sporulation and secondary infection of the rice blast fungus, *Magnaporthe oryzae. Pest. Manag. Sci.* 69:278-284.

[66] Kunova, A., et al. 2013. Sensitivity of non-exposed and exposed populations of *Magnaporthe oryzae* from rice to tricyclazole and azoxystrobin. *Plant Dis: accepted for pubblication.*

[67] Kurahashi, Y. 2001. Melanin biosynthesis inhibitors (MBIs) for control of rice blast. *Pestic Outlook* 12.32-35.

[68] Kurschner, E., et al. 1992. Effects of nitrogen timing and split application on blast disease in upland rice. *Plant Dis.* 76:384-389.

[69] Liu, W., et al. 2011. Multiple plant surface signals are sensed by different mechanisms in the rice blast fungus for appressorium formation. *PLoS Pathog.* 7:e1001261.

[70] Long, D. H., et al. 2001. Rice blast epidemics initiated by infested rice grain on the soil surface. *Plant Dis.* 85:612-616.

[71] Long, D. H., et al. 2000. Effect of nitrogen fertilization on disease progress of rice blast on susceptible and resistant cultivars. *Plant Dis.* 84:403-409.

[72] Ma, B., et al. 2009. Baseline and non-baseline sensitivity of *Magnaporthe oryzae* isolates from perennial ryegrass to azoxystrobin in the northeastern United States. *Can. J. Plant Pathol.* 31:57-64.

[73] Manandhar, H. K., et al. 1998. Seedborne infection of rice by *Pyricularia oryzae* and its transmission to seedlings. *Plant Dis.* 82:1093-1099.

[74] Manosalva, P. M., et al. 2009. A Germin-Like Protein Gene Family Functions as a Complex Quantitative Trait Locus Conferring Broad-Spectrum Disease Resistance in Rice. *Plant Physiol.* 149:286-296.

[75] Marcel, S., et al. 2010. Tissue-adapted invasion strategies of the rice blast fungus *Magnaporthe oryzae*. *Plant Cell* 22:3177-3187.

[76] Markham, P., et al. 1993. Choline - its role in the growth of filamentous fungi and the regulation of mycelial morphology. *Fems Microbiol. Rev.* 104:287-300.

[77] Meung, H., et al. 2003. Using genetic diversity to achieve sustainable rice disease management. *Plant Dis.* 87:1156-1169.

[78] Misato, T. 1967. Blasticidin S. Pages 434-439 In: *Antibiotics. Mechanism of Action,* vol. 1. D. Gottlieb and P. Shaw, eds. Springer Berlin Heidelberg.

[79] Molina, J., et al. 2011. Molecular evidence for a single evolutionary origin of domesticated rice. *Proceedings of the National Academy of Sciences* 108:8351-8356.

[80] Morinaga, H., et al. 2004. A benzimidazole fungicide, benomyl, and its metabolite, carbendazim, induce aromatase activity in a human ovarian granulose-like tumor cell line (KGN). *Endocrinology* 145:1860-1869.

[81] Morton, V. and Staub, T. 2008. A short history of fungicides. In: *Online. APSnet Features.* doi: 10.1094/APSnetFeature-2008-0308.

[82] Mueller, D. 2006. Fungicides: Triazoles. *Integrated Crop Management* 469:150-151.

[83] Nelson, R., et al. 2001. Working with resource-poor farmers to manage plant diseases. *Plant Dis.* 85:684-695.

[84] Okuno, T., et al. 1983. Mechanism of inhibitory effect of tricyclazole on secondary infection by spores of *Pyricularia oryzae. J. Pestic Sci.* 8:361-362.

[85] Ou, S. H. 1985. Blast - *Pyricularia oryzae.* Pages 109-210 In: *Rice Diseases.* S. H. Ou, ed. Cambrian News Ltd., Aberystwyth (Great Britain).

[86] Ou, S. H. 1980. A look at worldwide rice blast disease control. *Plant Dis.* 64:439-445.

[87] Paplomatas, E. J., et al. 2005. Molecular characterization and biological response to respiration inhibitors of *Pyricularia* isolates from ctenanthe and rice plants. *Pest. Manag. Sci.* 61:691-698.

[88] Prabhu, A. S., et al. 2003. Cultivar response to fungicide application in relation to rice blast control, productivity and sustainability. *Pesquisa Agropecuária Brasileira* 38:11-17.

[89] Qamruzzaman and Nasar, A. 2013. Degradation of tricyclazole by colloidal manganese dioxide in the absence and presence of surfactants. *Journal of Industrial and Engineering Chemistry.*

[90] Qu, S. H., et al. 2006. The broad-spectrum blast resistance gene Pi9 encodes a nucleotide-binding site-leucine-rich repeat protein and is a member of a multigene family in rice. *Genetics* 172:1901-1914.

[91] Ramalingam, J., et al. 2003. Candidate Defense genes from rice, barley, and maize and their association with qualitative and quantitative resistance in rice. *Mol. Plant Microbe In.* 16:14-24.

[92] Rossaro, B., et al. 2013. The effects of tricyclazole treatment on aquatic invertebrates in a rice paddy field. CLEAN – Soil, Air, Water:n/a-n/a.

[93] Russell, P. E. 2005. A century of fungicide evolution. *J. Agr. Sci.* 143:11-25.

[94] Saleh, D., et al. 2012. Sex at the origin: an Asian population of the rice blast fungus *Magnaporthe oryzae* reproduces sexually. *Mol. Ecol.* 21:1330-1344.

[95] Sasaki, M. 2003. Fungicides, organophosphorus compounds. Pages 601-609 In: *Encyclopedia of Agrochemicals.* John Wiley and Sons, Inc.

[96] Sato, K., et al. 2001. QTL analysis of resistance to the rice blast pathogen in barley (*Hordeum vulgare*). *Theor. Appl. Genet.* 102:916-920.

[97] Savary, S., et al. 2000. Rice pest constraints in tropical Asia: Quantification of yield losses due to rice pests in a range of production situations. *Plant Dis.* 84:357-369.

[98] Savary, S., et al. 2000. Rice pest constraints in tropical Asia: Characterization of injury profiles in relation to production situations. *Plant Dis.* 84:341-356.

[99] Schenker, S. 2012. An overview of the role of rice in the UK diet. *Nutrition Bulletin* 37:309-323.

[100] Sesma, A. and Osbourn, A. E. 2004. The rice leaf blast pathogen undergoes developmental processes typical of root-infecting fungi. *Nature* 431:582-586.

[101] Shanmugam, T. R., et al. 2006. Quantification and prioritization of constraints causing yield loss in rice (*Oryza sativa*) in India. *Agricultura Tropica et Subtropica* 39:194-204.

[102] Sierotzki, H., et al. 2007. Cytochrome b gene sequence and structure of *Pyrenophora teres* and *P-tritici-repentis* and implications for QoI resistance. *Pest. Manag. Sci.* 63:225-233.

[103] Sierotzki, H., et al. 2000. Point mutation in cytochrome b gene conferring resistance to strobilurin fungicides in *Erysiphe graminis* f. sp *tritici* field isolates. *Pesticide Biochemistry and Physiology* 68:107-112.

[104] Skamnioti, P. and Gurr, S. J. 2009. Against the grain: Safeguarding rice from rice blast disease. *Trends Biotechnol.* 27:141-150.

[105] Smith, C. M. 1988. History of benzimidazole use and resistance. Pages 41-43 In: *Fungicide resistance in North America.* C. J. Delp, ed. American Phytopathological Society.

[106] Staub, T., et al. 2003. Chemical activators of disease resistance. Pages 272-280 In: *Encyclopedia of Agrochemicals.* John Wiley and Sons, Inc.

[107] Sugiura, H., et al. 1993. Mutual antagonism between sterol demethylation inhibitors and phosphorothiolate fungicides on *Pyricularia oryzae* and the implications for their mode of action. *Pesticide Science* 39:193-198.

[108] Sundin, G. W. 2003. Antibiotics. In: *Encyclopedia of agrochemicals.* John Wiley and Sons, Inc.

[109] Suzuki, H. 1975. Meteorological factors in the epidemiology of rice blast. *Annu. Rev. Phytopathol.* 13:239-256.

[110] Syngenta. 2013. Amistar. http://www.syngenta.com/global/corporate/en/products-and-innovation/product-brands/crop-protection/fungicides/pages/amistar.aspx, accessed on 21/06/2013.

[111] Takagaki, M., et al. 2004. Mechanism of resistance to carpropamid in *Magnaporthe grisea. Pest. Manag. Sci.* 60:921-926.

[112] Talbot, N. J. 2003. On the trail of a cereal killer: Exploring the biology of *Magnaporthe grisea. Annu. Rev. Microbiol.* 57:177-202.

[113] Teng, P. S., et al. 1991. An analysis of the blast pathosystem to guide modeling and forecasting. Pages 1-30 In: *Rice Blast Modeling and Forecasting.* International Rice Research Institute (IRRI), Los Banos, Philipines.

[114] Teng, P. S., et al. 1990. Current knowledge on crop losses in tropical rice. Pages 39-53 In: *Crop loss assessment in rice.* IRRI (International Rice Research Institute).

[115] Thompson, J. E., et al. 1997. Trihydroxynaphthalene reductase from *Magnaporthe grisea*: Realization of an active center inhibitor and elucidation of the kinetic mechanism. *Biochemistry-Us* 36:1852-1860.

[116] Thompson, J. E., et al. 2000. The second naphthol reductase of fungal melanin biosynthesis in *Magnaporthe grisea. J. Biol. Chem.* 275:34867-34872.

[117] Uehara, T., et al. 1995. Effect of pyroquilon, an inhibitor of melanin synthesis, on sporulation and secondary infection of *Magnaporthe grisea. J. Phytopathol.* 143:573-576.

[118] Uesugi, Y. 2001. Fungal choline biosynthesis - a target for controlling rice blast. *Pesticide Outlook* 12:26-27.

[119] Urashima, A. S., et al. 2004. Resistance spectra of wheat cultivars and virulence diversity of *Magnaporthe grisea* isolates in Brazil. *Fitopatologia Brasileira* 29:511-518.

[120] Vincelli, P. and Dixon, E. 2002. Resistance to Q(o)I (strobilurin-like) fungicides in isolates of *Pyricularia grisea* from perennial ryegrass. *Plant Dis.* 86:235-240.

[121] Webster, R. K. and Gunnell, P. S. 1992. *Compendium of rice diseases.* APS Press.

[122] Wilson, R. A. and Talbot, N. J. 2009. Under pressure: Investigating the biology of plant infection by *Magnaporthe oryzae. Nat. Rev. Microbiol.* 7:185-195.

[123] Woloshuk, C. P., et al. 1980. Melanin biosynthesis in *Pyricularia oryzae*: Site of tricyclazole inhibition and pathogenicity of melanin-deficient mutants. *Pestic Biochem. Physiol.* 14:256-264.

[124] Wood, P. M. and Hollomon, D. W. 2003. A critical evaluation of the role of alternative oxidase in the performance of strobilurin and related

fungicides acting at the Q(o) site of Complex III. *Pest. Manag. Sci.* 59: 499-511.

[125] Yamaguchi, J., et al. 2002. Decreased effect of carpropamid for rice blast control in the west north area of Saga Prefecture in 2001. *Jpn. J. Phytopathol.* 68:261.

[126] Yang, J., et al. 2009. Expression and homology modelling of sterol 14α-demethylase of *Magnaporthe grisea* and its interaction with azoles. *Pest. Manag. Sci.* 65:260-265.

[127] Yudelman, M., et al. 1998. *Pest management and food production. Looking to the future.* International Food Policy Research Institute, Washington.

[128] Yukioka, H., et al. 1998. Transcriptional activation of the alternative oxidase gene of the fungus *Magnaporthe grisea* by a respiratory-inhibiting fungicide and hydrogen peroxide. *Bba-Gene Struct. Expr.* 1442:161-169.

[129] Zeigler, R. S. 1998. Recombination in *Magnaporthe grisea. Annu. Rev. Phytopathol.* 36:249-275.

[130] Zhang, C.-Q., et al. 2009. Resistance development in rice blast disease caused by *Magnaporthe grisea* to tricyclazole. *Pestic Biochem. Physiol.* 94:43-47.

[131] Zhang, C. Q., et al. 2005. Sensitivity detection technique and resistance risk assessment of *Magnaporthe grisea* to tricyclazole. *Rice Science* 12: 68-74.

[132] Zhang, C. Q., et al. 2006. Isolation, characterization and preliminary genetic analysis of laboratory tricyclazole-resistant mutants of the rice blast fungus, *Magnaporthe grisea. Journal of Phytopathology* 154:392-397.

[133] Zhu, Y., et al. 2000. Genetic diversity and disease control in rice. *Nature* 406:718-722.

[134] Zou, X.-H., et al. 2013. Multilocus estimation of divergence times and ancestral effective population sizes of *Oryza* species and implications for the rapid diversification of the genus. *New Phytologist* 198:1155-1164.

Chapter 3

FUNGICIDES AND THEIR ROLE IN DISEASE MANAGEMENT

Agostino Santomauro and *Franco Faretra*

Department of Plant, Soil, and Food Sciences,
University of Bari "Aldo Moro", Bari, Italy

ABSTRACT

Fungal diseases are known to affect produce quality on most crops and to cause severe economic losses worldwide. The use of fungicides has been for a long time the main tool to control many economically important fungal pathogens. In the last decades, increased public concern on food safety and environmental impact of farming activities has led regulatory Authorities to adopt very stringent rules on the evaluation and authorization of plant protection products (PPPs), especially for toxicological and environmental negative side effects related to their use. As a result, the development of new PPPs has been addressed at reducing their toxicity towards humans, mammalians and non-target species. In this framework, agrochemical firms have made available several new fungicides with novel specific (single-site) modes of action. Many of them have penetrant properties, are very selective, showing specific activity against target pathogens, as well as enhanced efficacy as compared to traditional multisite fungicides (i.e., carbamates, copper compounds, dithiocarbamates, quinones, sulphur, thiophtalimides, etc.).

---

* Corresponding Author: agostino.santomauro@uniba.it.

On the other hand, their single-site modes of action make them at moderate to high risk of inducing fungicide resistance in target pathogens and, hence, anti-resistance strategies must be carefully implemented in their use. Nowadays, about 200 active substances are available, with almost 40 different known specific target sites, plus a number of chemicals having multi-site or unknown mode of action. The increased availability of fungicides allows a more flexible and effective planning of crop protection strategies well fitting to Integrated Pest Management (IPM) principles. Yet, it requires a constant update of technical and scientific expertise by farmers and advisors. This review focuses on the main features of the most commonly used fungicides and their role in disease control in the overall approach to IPM aiming at improving cropping sustainability.

OVERVIEW ON FUNGICIDES' HISTORY AND EVOLUTION

Human's history is strictly related on his dependence on food plants. The survival of thousands of people has been endangered, along the centuries, by famines occurred as a consequence of plant diseases, as mentioned in the Bible, in writings of Latin and Greeks or as documented by historical epidemics like the "Irish famine", caused by *Phytophthora infestans* on potato in mid 1840s, the brown spot of rice, known as the "Bengal famine" (1943), and others (Belli, 2012).

Up even to the early 1800s, the lack of knowledge about the causal agents of plant diseases, ascribed to supernatural causes, hampered their control and, even after the discovery of the compound microscope (mid 1600s) that allowed to identify and study several fungal structures, the scientific community still continued to strongly believe that microorganisms and their spores were the result and not the cause of the diseases, according to the "spontaneous generation" theory (Agrios, 2005; Kelman and Peterson, 2002).

It was not until early 1860s that Louis Pasteur finally proved the real cause/effect relationship between microorganisms and diseases, even if some important deductions on this way had already been argued on wheat smut by Tillet in 1755 and Prévost in 1807, as well as on potato blight by Berkeley in 1845 and deBary in 1861. Nonetheless, the properties of sulfur in somehow controlling some plant injuries had been known for centuries and other concoctions like salt water and lime were used as well to treat wheat seeds against bunt (Maude, 1996). Lime sulfur was mentioned to be used on peach against powdery mildew (McCallan, 1967) and sulfur was successfully employed as a fungicide in France in the mid 1850s for the protection of grapevine from powdery mildew (Drandarevski et al., 1978), disease introduced from America in the 1840s that affected vines reducing wine production by 80% in 1854.

The first milestone in the fungicides' discovering history is, however, to be accredited to the French botany Pierre Alexis Millardet who, in 1885, occasionally observed that vines sprayed with a mixture of copper sulfate and lime were not affected by the disease that was destroying neighboring untreated plants. The Bordeaux mixture was born and copper-based compounds are still nowadays largely used throughout the world for the control of different agents of downy mildew on several crops followed, in early 1900s, by the development of organic mercurials for seed treatments and derivatives of arsenics.

The era of organic fungicides conventionally starts in 1940 with the commercialization of thiram, the first dithiocarbamate, a group of broad-spectrum organic fungicides, followed by the ethylenebisdithiocarbamates (mancozeb, maneb, zineb), all providing satisfactory control of several pathogens causing severe diseases, like potato blight, apple scab and leaf spot on several crop plants (Russell, 2005). The main fungicides discovered or commercialized up to 1960s and their main features are summarized in Table 1.

A further major step in fungicides' history was the discovery, in late 1960s, of carboxin, the first systemic fungicide used for seed treatments, and benzimidazoles, the first group of general purpose chemicals with systemic properties. As from that time, a number of new systemic fungicides with different specific target sites have been made available by agro-chemical companies. A non-exhaustive list of such molecules and their main features is summarized in Table 2.

The discovery of systemic fungicides represented a breakthrough in the perspective of plant disease managing since, as compared to non-systemic ones, they could offer several advantages like: redistribution into the plant and potential curative activity, rain-fastness, enhanced selectivity, efficacy at lower dosage rates. On the other hand, their single-site modes of action make them at moderate to high risk of inducing fungicide resistance in target pathogens. Such a trouble was first experienced at a large scale with benzimidazoles soon after their introduction, in early 1970s (Bollen and Scholten, 1971).

Table 1. Main fungicides discovered or commercialized up to 1970s and their main features

Years[1]	Group name[2]	Chemical group[2]	ISO names of fungicides	Spectrum of activity	Target site (mobility)	Risk of resistance[2]
1800s	Inorganic	Inorganic	Sulphur	Mainly powdery mildews	Multi site (covering)	Low
1880s	Inorganic	Inorganic	Copper (salts)	Downy mildews	Multi site (covering)	Low
1930s-1960s	Dithiocarbamates and ethylenebis dithiocarbamates	Dithiocarbamates and ethylenebis dithiocarbamates	Ferbam Mancozeb Maneb Metiram Propineb Thiram Zineb Ziram	Broad spectrum	Multi site (covering)	Low
1940s-1950s	Dinitrophenyl Crotonates	Dinitrophenyl crotonates	Dinocap	Powdery mildews	Uncouplers of oxidative phosphorylation (covering)	Low
1950s	Guanidines	Guanidines	Dodine	Broad spectrum	Not known	Low to medium
1950s-1960s	Phtalimides	Phtalimides	Captan Captafol Folpet	Broad spectrum; Seed treatments	Multi site (covering)	Low
1960s	Quinones	Quinones	Dithianon	Broad spectrum	Multi site (covering)	Low
1960s	Chloronitriles	Chloronitriles	Chlorothalonil	Broad spectrum	Multi site (covering)	Low
1960s	Succinate dehydrogenase	Carboxamides	Carboxin Oxycarboxin	Basidiomycota (smuts, rusts)	Respiration at complex II level	Medium to high

Years[1]	Group name[2]	Chemical group[2]	ISO names of fungicides	Spectrum of activity	Target site (mobility)	Risk of resistance[2]
	inhibitors (SDHI)				(penetrant)	
1950s-1970s	Methyl Benzimidazole Carbamates (MBC)	Benzimidazoles	Benomyl Carbendazim Thiabendazole	Broad spectrum (no activity on downy mildews)	ß-tubuline assembly in mitosis (penetrant)	High
		Thiophanates	Thiophanate-methyl			

[1] Years of discovery or commercialization.
[2] As specified in FRAC Code List, 2013 (www.frac.info).

Table 2. Main fungicides discovered or commercialized from 1970s onwards and their main features

Group name[1]	Chemical group[1]	ISO names of fungicides	Spectrum of activity	Target site[1] (mobility)	Risk of resistance[1]
Cyanoacetamide oxime	Cyanoacetamide oxime	Cymoxanil	Downy mildews	Not known (penetrant)	Low to medium
Phosphonates	Ethyl phosphonates	Fosetyl-Aluminium	Downy mildews	Not known (penetrant)	Low
Dicarboximides	Dicarboximides	Iprodione Procymidone Vinclozolin	*Botrytis, Sclerotinia, Monilia* spp.	Histidine- Kinase in osmotic signal transduction (penetrant)	Medium to high
Amines (SBI: Class II)	Morpholines	Dodemorph Tridemorph	Powdery mildews	$\Delta 14$ reductase and $\Delta 8$- $\Delta 7$ isomerase in sterol biosynthesis (penetrant)	Low to medium
	Piperidines	Fenpropidin	Powdery mildews, rusts on cereals		
	Spiroketalamines	Spiroxamine	Powdery mildews		

Table 2. (Continued)

Group name[1]	Chemical group[1]	ISO names of fungicides	Spectrum of activity	Target site[1] (mobility)	Risk of resistance[1]
Demethylation Inhibitors (DMI) (SBI: Class I)	Piperazines	Triforine	Powdery mildews, rusts	C14 demethylase in sterol biosynthesis (penetrant)	Medium
	Imidazoles	Imazalil Prochloraz	Broad spectrum for seed treatments (imazalil also post harvest)		
	Pyrimidines	Fenarimol	Powdery mildews, scab		
	Triazoles	Bitertanol Cyproconazole Hexaconazole Myclobutanil Penconazole Tebuconazole Tetraconazole Triadimenol	Powdery mildews, scab, *Monilia* spp.,		
(SBI: Class III)	Hydroxyanilides	Fenhexamid	*Botrytis, Sclerotinia, Monilia* spp.	3 ketoreductase, C4 demethylation (covering)	Low to medium
	Aminopyrazolinone	Fenpyrazamine	*Botrytis* spp.		
Phenylamides	Acylalanines	Benalaxyl Benalaxyl-M Metalaxyl Metalaxyl-M	Downy mildews	RNA polymerase I (penetrant)	High
	Oxazolidinones	Oxadixyl			
Hydroxypyrimidines	hydroxypyrimidines	Bupirimate	Powdery mildews	adenosin-deaminase (penetrant)	Medium
Benzamides	Toluamides	Zoxamide	Powdery mildews	ß-tubulin assembly in mitosis (covering)	Low to medium

Group name[1]	Chemical group[1]	ISO names of fungicides	Spectrum of activity	Target site[1] (mobility)	Risk of resistance[1]
Benzamides	pyridinylmethyl-benzamides	Fluopicolide	Downy mildews	delocalisation of spectrin-like proteins (penetrant)	Not known
Succinate dehydrogenase inhibitors (SDHI)	Pyridine carboxamides	Boscalid	*Botrytis cinerea, Erysiphe necator, Stemphylium vesicarium, Sclerotinia, Alternaria* and *Monilia* spp.	Respiration at complex II level (covering/penetrant)	Medium to high
	Pyridinyl ethyl benzamides	Fluopyram	Broad spectrum; mainly, *Botrytis* and powdery mildews	Respiration at complex II level (penetrant)	
Quinone outside Inhibitors (QoI)	Methoxy acrylates	Azoxystrobin Picoxystrobin	Broad spectrum	Respiration at complex III: cytochrome bc1 at Qo site (covering/penetrant)	High
	Methoxy carbamates	Pyraclostrobin	Broad spectrum		
	oximino acetates	Kresoxim-methyl Trifloxystrobin	Broad spectrum, mainly powdery mildews		
	Oxazolidine diones	Famoxadone	Broad spectrum, mainly downy mildews		
	Dihydro-dioxazines	Fluoxastrobin	Broad spectrum, seed treatments		
	Imidazolinones	Fenamidone	Broad spectrum, mainly downy mildews		
Quinone inside Inhibitors (QiI)	Cyano- imidazole	Cyazofamid	Downy mildews	Respiration at complex III: cytochrome bc1 at Qi site (penetrant)	Medium to high
	Sulfamoyl-triazole	Amisulbron	Downy mildews		
Quinone x Inhibitors (QxI)	Triazolo pyrimidylamine	Ametoctradin	Downy mildews	Respiration at complex III: cytochrome bc1 at Qx unknown site (covering)	Medium to high

Table 2. (Continued)

Group name[1]	Chemical group[1]	ISO names of fungicides	Spectrum of activity	Target site[1] (mobility)	Risk of resistance[1]
Anilino Pyrimidines	Anilino Pyrimidines	Cyprodinil Mepanipyrim Pyrimethanil	Scab, Monilia Botrytis,	Methionine Biosynthesis (penetrant)	Medium
Azanaphthalenes	Aryloxyquinoline	Quinoxyfen	Powdery mildews	Not known (covering/penetrant)	Medium
	Quinazolinone	Proquinazid	Powdery mildews		
Phenylpyrroles	Phenylpyrroles	Fludioxonil	Broad spectrum, *Botrytis*, seed treatments, post harvest	Histidine Kinase in signal transduction (covering)	Low to medium
Carbamates	Carbamates	Propamocarb	Downy mildews	Permeability of cell membrane (penetrant)	Low to medium
Carboxylic Acid Amides (CAA)	Cinnamic acid amides	Dimethomorph	Downy mildews	Cellulose synthase (penetrant)	Low to medium
	Valinamide carbamates	Benthiavalicarb Iprovalicarb Valifenalate	Downy mildews		
	Mandelic acid amides	Mandipropamid	Downy mildews		
Phenyl acetamide	Phenyl acetamide	Cyflufenamid	Powdery mildews	Not known (penetrant)	Resistance known in *Sphaerotheca fusca*
Aryl phenyl ketones	benzophenone	Metrafenone	Powdery mildews	Not known (penetrant)	Medium
	benzoylpyridine	Pyriofenone			

[1] As specified in FRAC Code List, 2013 (www.frac.info).

MODES OF ACTION AND RESISTANCE TO FUNGICIDES

The mode of action and risk of resistance of most commonly available fungicides are reported in Tables 1 & 2.

Spontaneous mutations, naturally occurring in fungal populations, can include the onset of resistance to a given fungicide at low frequency. It is only under the selection pressure operated by repeated applications of the fungicides with a same mode of action that the fungal population ratio shifts to resistant mutants leading, over time, to the reduction or even the loss of fungicide's efficacy, since mutations are transmissible to progenies.

Genetic bases of fungal resistance towards different groups of fungicides have been widely investigated for several pathogens in the past years, through both classical and molecular tools (e.g., Brent and Hollomon, 2007a, 2007b; De Guido et al., 2007; De Miccolis et al., 2010; Faretra and Pollastro, 1991; Georgopoulos and Skylakakis, 1986; Gisi et al., 2002; Grindle and Faretra, 1993; Steffens et al., 1996; Thind, 2012; Zhonghua and Michailides, 2005).

Resistance can be triggered by either major or minor genes. The mutation of one or more major genes causes a qualitative change of the fungal phenotype, with a discrete increase of the level of resistance (qualitative or discrete resistance). The stable loss of efficacy of benzimidazoles towards many pathogens is an example of this kind of resistance.

The mutation of several minor genes, each causing just a little change in the level of resistance is usually indicated as quantitative or continuous resistance. It occurs very gradually, with variable degrees in population sensitivity and, as such, is difficult to appreciate in the field, even if high frequencies of partially resistant individuals can be present in fungal populations. This is the case of resistance to DMIs.

The cross-resistance is the phenomenon for which once a pathogen acquires resistance to a given fungicide, it becomes resistant also to other ones with the same specific (single-site) mode of action. Cross-resistance is present within groups of fungicides like anilino-pyrimidines, CAAs, dicarboximides, DMIs, phenylamides, QoIs, SDHIs.

A different situation, known as multiple resistance, is given by the occurrence of independent mutations causing the concurrent resistance of a fungus to fungicides having different modes of action. For instance, strains of *Botrytis cinerea* simultaneously resistant up to six different groups of fungicides were recently detected on table grape and strawberry in Southern Italy (De Miccolis Angelini et al., unpublished data). Simultaneous resistance to more unrelated fungicides can also be ascribed to the so called multi-drug resistance (MDR) mechanism, as described below.

From a technical and scientific point of view, the onset of mutations leading to any reduction in sensitivity to fungicides is intended as resistance, regardless of its occurring under experimental conditions or in the field. From a practical point of view, only the reduction or failure of disease control in the field, as a consequence of fungicide applications should be referred to as practical or field resistance (Brent and Hollomon, 2007a).

The overall risk of resistance depends on a combination of the multi-site/single-site mode of action of the concerned fungicide and pathogen's potential to develop resistant mutants, as well as cropping conditions (Brent and Hollomon, 2007a). The situations at the highest risk are those having the following conditions: repeated sprays with fungicides having a same single-site mode of action; pathogens with abundant sporulation and short biological cycle causing polycyclic diseases; cropping under glass houses, reducing or preventing the competition with normally sensitive strains; resistant mutants with high fitness; monogenic determinism of resistance, since mutations in major genes are more probable to occur and are easier to cause the loss of fungicide efficacy under field conditions.

The resistance risk analysis is nowadays one of the key issues to compulsory include in the dossier for the official registration of fungicides. Yet, even after product's authorization and commercialization, there is the need to monitor and prevent the eventual occurrence of resistance, at least during early years of its use. Such need led agro-chemical companies to constitute the Fungicide Resistance Action Committee (FRAC), also aimed at establishing resistance management guidelines to prolong the effectiveness of "at risk" fungicides (www.frac.info). A monograph on the assessment of risk of fungicide resistance was prepared by the FRAC (Brent and Hollomon, 2007b) and a specific guideline [Standard PP 1/213 (3)] to this regard was made available by the European and Mediterranean Plant Protection Organization (EPPO; www.eppo.int).

RESISTANCE MECHANISMS

Mutations giving rise to resistance cause biochemical and physiological modifications within fungal cells (e.g., Dekker, 1976; Ma and Michailides, 2005). The most frequent are: i) decrease of the affinity of the site of action to the fungicide: this is, for instance, the case of lost of affinity of β-tubulin to benzimidazoles (Davidse, 1986) and of DNA-dependent RNA polymerase I to phenylamides (Davidse, 1988); ii) reduction of membranes permeability or mechanisms of extrusion of the fungicide known as multi-drug resistance (MDR); iii) detoxification or lack of conversion to toxic compound of the fungicide; iv) compensation of the fungicide activity by, for instance, an increased production of the target enzyme; and v) activation of metabolic pathways alternative to that inhibited by the fungicide as it occurs, for instance, with the activation of an alternative oxidase (AOX) overcoming the inhibition of respiration by QoI fungicides (Miguez et al., 2004; Wood and Hollomon, 2003).

MDR is a mechanism based on active export of toxic compounds from fungal hyphae, mediated by plasma membrane transporters. In fungi, ATP Binding Cassette (ABC) transporters and Major Facilitator Superfamily (MFS) transporters have been identified to be mainly responsible of MDR. Such mechanism have been thoroughly studied in different resistant phenotypes of *B. fuckeliana*, where the increase of fungicide efflux due to ABC transporters gives rise to MDR1, MFS to MDR2 and the concurrent activity of both transporters is responsible of MDR3 (Leroux and Walker, 2013). MDR mechanism is documented to occur also in other fungi, like *Mycosphaerella graminicola* and *Penicillium digitatum* (De Waard et al., 2006; Gupta and Chattoo, 2008, Hamamoto et al., 2001; Kretschmer, 2012; Kretschmer et al., 2009; Leroux et al., 2010).

CURRENT GLOBAL SCENARIO

Nowadays, food security still constitutes a priority for millions of people worldwide. According to the FAO 2012 report on the State of Food Insecurity in the World, about 870 million people are undernourished, most of all living in developing Countries like Asia, Latin America and Africa, but even in developed regions, the number of undernourished people has increased from 13 million in 2004-2006 to 16 million in 2010-2012 (FAO, 2012).

Food availability at global level is mainly threatened by population growth, envisaged to increase from 6.8 billion in 2010 to 9.2 and 9.9 billion by 2050 and 2100, respectively (United Nations, 2010) and by the reduction of productive land areas, foreseen to 0.18 ha per capita by 2050, from 0.25 in 2000 (Alexandratos and Bruinsma, 2012), due to several concurrent forces like industrialization, urbanization (markedly in developing Countries), as well as growing demand for renewable energies resulting in subtraction of agricultural land for the cultivation of bio-fuel crops, even if several uncertainties hamper any sound projection on the future spread of such crops (Alexandratos and Bruinsma, 2012).

The FAO report underlines also the need of further agricultural growth to reduce hunger and malnutrition in poor Countries, along with the increase of food quality in terms of diversity, nutrient content and safety. Concern on food safety and environmental impact of farming activities has progressively grown also in the developed Countries in the last decades, leading to a general demand for a "safe" agriculture, particularly focusing on the reduction or, even, the suppression of chemical inputs.

The perspective of an increase of both the quantity and quality of food by simultaneously producing in an environmental-friendly manner in the forthcoming decades, is a major challenge to face in a multidisciplinary approach, involving different expertise in many fields, as biology, toxicology and eco-toxicology, environment, agronomy, agro-meteorology, genetics and chemistry.

In the sector of plant diseases, such need strongly stimulated, during the last years, public and private research in finding alternative/safer means and measures for the control of pathogens and regulatory Authorities in adopting more and more stringent rules on the authorization of plant protection products (PPPs), especially with respect to toxicological and environmental negative side effects related to their use (see below). This resulted in the development and commercialization of bio-fungicides (microbial antagonists) and of new chemicals characterized by reduced toxicity towards humans, mammalians and non-target species, novel modes of action, specific activity against target pathogens, and enhanced efficacy at lower dosage rates as compared to traditional fungicides.

It is out of doubt that, since no more land would be available, the increase of crop yields would be necessarily gained through the improvement of productivity and, in such a context, the correct and rationale use of fungicides would be therefore expected to play a key role in effectively limiting yield losses caused by plant diseases, especially in developing Countries, as well as by alien pathogens introduced into new areas, via the global trading and mass tourism (McGee, 1997; Vurro et al., 2010). Moreover, it is not of secondary importance their contribution to food safety and quality by preventing food and feed contamination by mycotoxins, secondary fungal metabolites toxic to animal and humans, produced mainly by *Aspergillus*, *Penicillium* and *Fusarium* genera, representing a serious concern at global level and, markedly, in developing Countries of Africa, Asia and South America (Bennet and Klich, 2003; Berthiller et al, 2013; Moss, 2002a, 2002b; Richard, 2007).

Reliable and updated information on the percentages of crop losses caused by fungal diseases is not easy to be gathered due to a number of variables, like the considered crop and concerned geographic area, the lack of nets of monitoring and surveys in some Countries, with particular regard to developing ones, the lack of data related to minor, even important crops at local level, etc. Nevertheless, according to some major estimates, it could be assumed that losses due to diseases, insects, and weeds on major crops can potentially range from 30 to 40% worldwide, with an incidence of plant pathogens of 15-30%. (Gianessi and Reigner, 2006; Oerke, 2006; Oerke et al., 1994).

GENERAL OUTLINES ON THE PROCESS FOR AUTHORIZATION OF PLANT PROTECTION PRODUCTS IN THE EU

The EU legislation on PPPs is maybe one of the most stringent and articulated ones. It was in early 1990s that the issuing of the Directive 414/1991 regulated and harmonized the evaluation, marketing and use of PPPs (herbicides, insecticides, fungicides, etc.) at EU level. It established common rules for testing requirements and high common standards of safety and environmental stability and introduced a program for re-evaluating, with the harmonized criteria, all the active substances (a.s.) already existing in the market before July 1993. This resulted in the withdrawing of about 700 a.s. out of 920 existing a.s. at the end of the peer-reviewed process.

Such Directive has been recently repealed by the Regulation 1107/2009. The basic principles of the Directive remained unchanged, but new requirements and restrictions were introduced, like more tight deadlines for the evaluation processes, specific "cut-off" (hazard based) criteria for humans and the environment, the assessment of safeners and synergists like the a.s. and the introduction of the "low-risk" and "candidates for substitution" categories of a.s..

The EU legislation lays down a comprehensive risk assessment and approval procedure for a.s. and authorization of products containing such substances. Each a.s. has to be proven safe in terms of human health, including residues in the food chain, animal health and the environment, in order to be allowed to be approved. The first step of the evaluation process involves a Rapporteur Member State, which transmits its preliminary conclusions on the a.s. to the European Food Safety Authority (EFSA). A scientific risk assessment involving EFSA is then carried out, followed by risk management steps carried out by the EU Commission with the assistance of the Member States (MS) within the Standing Committee on the Food Chain and Animal Health. If the evaluation shows that the a.s. has no harmful effect on human or animal health and that it has no unacceptable influence on the environment it can be approved. A list of approved a.s. is established at EU level and MS may authorize only products containing such a.s.. According to the new Regulation, EU is divided into 3 zones: South, Central and North. Applications for the authorization of products are made at zonal level. For uses in glasshouses, seed treatment, post-harvest and storage stores EU is considered as one zone.

DISEASE MANAGEMENT

The concept of Integrated Pest Management (IPM), as from its first enunciation in early 1970s, in which it was addressed to the control of insect pests (Kogan, 1998), has over time evolved and extended to other harmful organisms, including fungal plant pathogens. Despite it sounds very familiar to all the people dealing with plant protection issues, IPM has been differently defined and implemented at various levels. In 2002, Bajwa and Kogan listed 67 proposed definitions of IPM, found in worldwide literature between 1959 and 2000 (Bajwa and Kogan, 2002). In recent years, the concept further evolved in Integrated Crop Management (ICM), Integrated Production or Sustainable Agriculture, in which IPM is the part focusing on disease management (Gisi et al., 2009). This results in a system approach integrating a variety of management techniques in a multidisciplinary view, aiming at the attainment of high quality agricultural produces in a sustainable way, by safeguarding the environment, the health of workers and consumers and the profitability of growers.

According to such principles, chemicals should be applied only when a pest or disease cannot be controlled by other means. To this aim, a constant and accurate monitoring should be executed in the field, to decide whether and when to carry out chemical sprays, also relying on thresholds defining the economically damaging levels, where available. Yet, for many fungal diseases/crop systems no thresholds are available and/or actually applicable. Moreover, even if agro-chemical companies often claim for curative activity of several fungicides, it is common experience that once first symptoms of a fungal disease appear it would be too late for its effective control, especially for produces for which the market requires high quality standards. That's why, in the case of fungal pathogens, the planning and setting of protection strategies must be often based on a prevention approach that means to apply fungicides in critical times in the season prior to the disease outbreak.

The possibility to predict if plant infection is likely to occur, would greatly help in deciding about the actual need to intervene or not, thus rationalizing control strategies and avoiding unnecessary fungicide sprays. Forecasting models of several plant diseases have been so far developed since the first proposed by Van der Plank (1960, 1963) and applied to various crops, including relevant ones as potato, cereals, rice, grapevine, etc. (for review see, e.g. Agrios, 2005; Contreras-Medina et al., 2009; De Wolf and Isard, 2007; Gessler et al., 2011; Hardwick, 2006; Pan et al., 2010; Shaw, 1994; Van Maanen and Xu, 2003). Some of them are simple empirical models based on the analysis of the relationships between epidemics development and the biological and meteorological parameters. Analyses can be based on empirical observations in the field or on statistical methods. Examples of such empirical models are the tables of Mills (1944) for the prediction of apple scab, providing information about the minimum hours of leaf wetness required for the infection as related to temperature, or the so called "Three 10 rule", developed by Baldacci (1947) predicting downy mildew infections on grapevine, based on the simultaneous occurrence of i) at least 10° C of temperature; ii) at least 10 cm in length of shoots; iii) at least 10 mm of rainfall in 24–48 h. A different kind of models are the mechanistic ones that are based on the analysis of the stages of pathogen's infection cycle, as influenced by host and environmental variables (De Wolf and Isard, 2007). One example of such models is the one recently developed in Italy known as UCSC, which simulates the downy mildew disease cycle on grapevine, from oospore maturation through the onset of symptoms (Caffi et al., 2007). Recently, a hybrid method integrating qualitative differential models and fuzzy-neural systems has been proposed for *Plasmopara viticola*, the causal agent of grapevine downy mildew (Vercesi et al., 2010).

Plant disease models, especially empirical ones, are often developed in specific climates with specific conditions that might not apply for all areas. Therefore, models normally need to be calibrated and validated in practice, for one or more seasons, to verify whether they are accurate and robust enough to correctly predict infections under local conditions

The recent Directive 2009/128/EC on the Sustainable Use of Pesticides makes the adoption of tools for pest monitoring and of Decision Supporting Systems (DSS), where available, as well as the implementation of the general principles of IPM mandatory by 1st January 2014 for all the Member States of the EU.

A key aspect in applying IPM strategies is the management of resistance. Many of the modern fungicides are at medium to high resistance risk and therefore a management strategy should be implemented before resistance becomes an economically important problem. An anti-resistance strategy must be a part of the overall IPM framework and, therefore, it must take into consideration the implementation of cultural practices aimed at preventing the outbreak of the diseases and/or at reducing fungal inoculum and disease pressure, such as crop rotations, sowing dates and densities, pruning and adoption of hygiene measures, use of resistant/tolerant cultivars where available, use of certified seeds and planting materials, use of balanced fertilization, and irrigation, etc.. Where available and effective, biological, physical and other non-chemical methods must be as well implemented and adapted time by time to specific pathosystems (pathogen/crop combinations).

As a general principle, the reduction of selection pressure should be gained through the alternation (establishing the maximum number of sprays per season) and/or the use of mixtures of fungicides with different modes of action. In particular, when preparing tank mixtures it would advisable to use a multi-site partner compound, where feasible, since a mixture of two single-site ones, even with different modes of action, would increase the risk of arising of double resistance, as compared to the separate use of the two fungicides (Brent and Hollomon, 2007a). Moreover, when available, it is preferable to use formulated mixtures.

The risk of fungicide resistance is a serious concern also from the point of view of agro-chemical Companies who are rightfully interested in preserving as much as possible the market life of their products. Therefore, limitations to the number of applications per season are nowadays clearly indicated on the labels of most of at-risk fungicides, to strictly abide, as well as, all the other label specifications.

CONCLUSION

The rapidly increasing globalization of markets and the consequent growing competition at international level is having some direct and indirect impact also on phytosanitary aspects and on crop protection management. Some growers' choices, driven by the need to differentiate their produces in terms of both timing and quality to accomplish consumer and market trends, result for instance in growing specific cultivars on the bases of their commercial features like earliness in ripening and quality characteristics but very seldom for their resistance or tolerance to diseases. It is not infrequent, moreover, the case of farmers unprepared for unexpected outbreak of diseases due to the introduction of plant propagating materials not certified or controlled for its phytosanitary status.

More and more growers are trading their produce through large-scale retailers that normally require the adoption of quality certification schemes by their suppliers. GLOBALGAP is maybe one of the most renowned certification protocol recognized at global level which needs, to be implemented, a high grade of managing and organizing capability of the farm at all steps of the production process. This represents a positive challenge for the growing of the agriculture sector, since it would further stimulate the development of a quality-based and sustainable production approach by the farmers.

On the other hand several large-scale retailers, with the aim of satisfying expectations of their customers, with particular regard to food safety, have developed proper binding production protocols foreseeing severe limitations on the number and residue level of active substances that can be found in supplied vegetables and fruits, often much lesser than those legally established. This is causing very serious problems to farmers since, to be in line with such protocols, they are led to repeatedly apply few molecules, often chosen on the basis of their residual behavior rather than of their rational integration in protection strategies, thus clearly disregarding all IPM and resistance management principles above discussed.

Such a commercial policy, clearly overstepping its boundaries, induces growers to wrong behaviors from a technical and scientific point of view, constituting a step back on the long and difficult way towards the reaching and completion of sustainable agriculture. It would be, therefore, advisable to give skilled technicians, scientific and institutional subjects and growers' organizations the possibility to discuss and review such strategies, having so a heavy impact on technical aspects of the crop protection management.

For what so far discussed, it comes evident how complex are the choices to be made in integrated production systems. In such an overall context, the use of fungicides, as properly planned and integrated with microbial antagonists, natural substances and other alternative means by professionally qualified consultants, transferring to the field updated information coming from the world of research and experimentation, is likely expected to be still of relevant importance in contributing to the improvement of food quality and yields in an environmental friendly manner, for present and future generations.

REFERENCES

Agrios G. N., 2005. Plant Pathology, fifth edition. Elsevier Academic Press, London, UK, pp. 922.

Alexandratos N. and Bruinsma J., 2012. World agriculture towards 2030/2050: the 2012 revision. *ESA Working Paper No. 12-03*, Agricultural Development Economics Division - Food and Agriculture Organization of the United Nations, pp. 147.

Bajwa, W. I. and Kogan M., 2002. Compendium of IPM Definitions. What is IPM and how is it defined in the Worldwide Literature? *IPPC Publication No. 998, Integrated Plant Protection Center (IPPC)*, Oregon State University, Corvallis, OR 97331, USA.

Baldacci E., 1947 Epifitie di *Plasmopara viticola* (1941–46) nell'Oltrepó Pavese ed adozione del calendario di incubazione come strumento di lotta. *Atti Istituto Botanico, Laboratorio Crittogamico* VIII, 45–85.

Belli G., 2012. Malattie delle piante che segnarono la storia. Altravista ed., Italy, pp. 136.

Bennet J. W. and Klich M., 2003. Mycotoxins. *Clinical Microbiology Reviews*, 16, 497-516.

Berthiller F., Crews C., Dall'Asta C., De Saeger S., Haesaert G., Karlovsky P., Oswald I.P., Seefelder W., Speijers G. and Stroka J., 2013. Masked mycotoxins: a review. *Molecular Nutrition and Food Research*, 57, 165-186.

Bollen G. J. and Scholten G., 1971. Acquired resistance to benomyl and some other systemic fungicides in a strain of *Botrytis cinerea* in cyclamen. *Netherland Journal of Plant Pathology*, 77, 83-90.

Brent K. J. and Hollomon D. W., 2007a. Fungicide resistance in crop pathogens: how can it be managed?. FRAC Monograph n. 1 (second, revised edition). Fungicide Resistance Action Committee, Brussels, Belgium, pp. 52.

Brent K. J. and Hollomon D. W., 2007b. Fungicide resistance: the assessment of risk. FRAC Monograph n. 2 (second, revised edition). Fungicide Resistance Action Committee, Brussels, Belgium, pp. 53.

Caffi T., Rossi V., Bugiani R., Spanna F., Flamini L., Cossu A. and Nigro C., 2009. Model predicting primary infections of *Plasmopara viticola* in different grapevine-growing areas of italy. *Journal of Plant Pathology*, 91, 535-548.

Contreras-Medina L. M., Torres-Pacheco I., Guevara-González R. G., Romero-Troncoso R. J., Terol-Villalobos I. R. and Osornio-Rios R. A., 2009. Mathematical modeling tendencies in plant pathology. *African Journal of Biotechnology*, 8, 7399-7408.

Davidse L. C., 1986. Benzimidazole fungicides: mechanism of action and biological impact. *Annual Review of Phytopathology*, 24, 43-65.

Davidse L. C., 1988. Phenylamide fungicides: mechanism of action and resistance. In: *Fungicide Resistance in North America* (Delp C., coord.), APS Press, St. Paul, USA, 63-65.

De Bary A., 1861. Ueber die Geschlechtsorgane von *Peronospora. Botanische Zeitung*, 19, 89–91.

De Guido M. A., De Miccolis Angelini R. M., Pollastro S., Santomauro A. and Faretra F., 2007. Selection and genetic analysis of laboratory mutants of *Botryotinia fuckeliana* resistant to fenhexamid. *Journal of Plant Pathology*, 89, 203-210.

Dekker J., 1976. Acquired Resistance to Fungicides. *Annual Review of Phytopathology*, 14, 405-428.

De Miccolis Angelini R. M., Habib W., Rotolo C., Pollastro S. and Faretra F., 2010. Selection, characterization and genetic analysis of laboratory mutants of *Botryotinia fuckeliana* (*Botrytis cinerea*) resistant to the fungicide boscalid. *European Journal of Plant Pathology*, 128, 185-199.

De Waard M. A., Andrade A. C., Hayashi K., Shoonbeck H. J, Stergiopoulos I. and Zwiers L. H., 2006. Impact of fungal drug transporters on fungicide sensitivity , multidrug resistance and virulence. *Pest Management Science*, 62, 195-207.

De Wolf E. D. and Isard S. A., 2007. Disease Cycle Approach to Plant Disease Prediction. *Annual Review of Phytopathology*, 45, 203-220.

Drandarevski C. A., Corke A. T. K., Jordan V. W. L., Sitterly W. R., Royle D. J., Wheeler B. E. J., Burchill R. T., Dixon G. R., Bulit J. and Lafon R., 1978. The powdery mildews. Spencer, D. M. ed., 565 pp.

FAO, 2012. The State of Food Insecurity in the World 2012 - Economic growth is necessary but not sufficient to accelerate reduction of hunger and malnutrition. Rome, FAO, 61 pp.

Faretra F. and Pollastro S., 1991. Genetic basis of resistance to benzimidazole and dicarboximide fungicides in *Botryotinia fuckeliana* (*Botrytis cinerea*). *Mycological Research*, 95, 943-951.

Georgopoulos S. G. and Skylakakis G., 1986. Genetic variability in the fungi and the problem of fungicide resistance. *Crop Protection*, 5, 299-305.

Gessler C., Pertot I., Perazzolli M., 2011. *Plasmopara viticola*: a review of knowledge on downy mildew of grapevine and effective disease management. *Phytopathologia Mediterranea*, 50, 3-44.

Gianessi, L. and Reigner, N., 2006. The importance of fungicides in U.S. crop production. *Outlook on Pest Management*, 10, 209-213.

Gisi U., Sierotzki H., Cook A. and McCaffery A., 2002. Mechanisms influencing the evolution of resistance to Qo inhibitor fungicides. *Pest Management Science*, 58, 859-867.

Gisi U., Chet I. and Gullino M. L, (eds.), 2009. Recent developments in management of plant diseases. Springer, UK, pp. 377.

Grindle M. and Faretra F., 1993. Genetic aspects of fungicide resistance. In: Lyr H. and Polter C. (eds.), Proceedings of the 10th International Symposium "Modern Fungicides and Antifungal Compounds". Ulmer, Wollgrasweg, Germany, 33-43.

Gupta A. and Chattoo B. B., 2008. Functional analysis of a novel ABC transporter ABC4 from *Magnaporte grisea*. *FEMS Microbiology Letters*, 278, 22-28.

Hamamoto H., Nawata O., Hasegawa K., Nakaune R., Lee Y.J., Makizumi Y., Akutsu K. and Hibi T., 2001. The role of the ABC transporter gene PMR1 in demethylation inhibitor resistance in Penicillium digitatum. *Pesticide Biochemistry and Phisiology,* 70, 19-26.

Hardwick N. V., 2006. Disease forecasting. In: *The epidemiology of plant diseases* 2nd edition, Cooke B. M., Gareth Jones D., Kaye B. (eds.), 239-267.

Kelman, A. and Peterson P. D., 2002. Contributions of plant scientists to the development of the germ theory of disease. *Microbes and Infection*, 4, 257-260.

Kogan M., 1998. Integrated pest management: historical perspectives and contemporary developments. *Annual Review of Entomology*, 43, 243–270.

Kretschmer M., 2012. Emergence of multi-drug resistance in fungal pathogens: a potential threat to fungicide performance in agriculture. In Thind T. S. (ed.): Fungicide resistance in crop protection. *Risk and management*, CAB International, UK, 2012, 251-267.

Kretschmer M., Leroch M., Mosbach A., Walker A. S., Fillinger S., Mernke D., Schoonbeek H. J., Pradier J. M., Leroux P., De Waard M. A. and Hahn M., 2009. Fungicide-driven evolution and molecular basis of multidrug resistance in field populations of the grey mould fungus *Botrytis cinerea*. *PLOS Pathogen*, 5, 12, e1000696.

Leroux P. and Walker A. S., 2013. Activity of fungicides and modulators of membrane drug transporters in field strains of *Botrytis cinerea* displaying multidrug resistance. *European Journal of Plant Pathology*, 135, 683-693.

Leroux P., Gredt M., Leroch M. and Walker A. S., 2010. Exploring mechanisms of resistance to respiratory inhibitors in field strains of *Botrytis cinerea*, the causal agent of gray mold. *Applied and Environmental Microbiology*, 76, 6615-6630.

Ma Z. and Michailides T. J., 2005. Advances in understanding molecular mechanisms of fungicide resistance and molecular detection of resistant genotypes in phytopathogenic fungi. *Crop Protection*, 24, 853-863.

Maude, R. B., 1996. Seedborne diseases and their control. CAB International, Wallingford, UK.

McCallan S. E. A., 1967. History of fungicides. In: Torgeson D.C. (ed.), Fungicides, Vol. I. Academic Press Inc., pp. 716.

McGee D. C., 1997. Plant pathogens and the worldwide movement of seeds. APS Press, Saint Paul, pp. 109.

Miguez, M., Reeve, C., Wood, P. M. and Hollomon, D. W., 2004. Alternative oxidase reduces the sensitivity of *Mycosphaerella graminicola* to QoI fungicides. *Pest Management Science*, 60, 3–7.

Mills W. D., 1944. Efficient use of sulfur dusts and sprays during rain to control apple scab. *Cornell Extension Bulletin*, 630, 4.

Moss M. O., 2002a. Micotoxin Review – 1. *Aspergillus* and *Penicillium*. *Mycologist*, 16, 116-119.

Moss M. O., 2002b. Micotoxin Review – 2. *Fusarium. Mycologist*, 16, 158-161.

Oerke E. C., 2006. Crop losses to pests. *The Journal of Agricultural Science*, 144, 31-43.

Oerke E. C., Dehne H. W., Schönbeck F. and Weber A., 1994. Crop Production and Crop Protection: Estimated Losses in Major Food and Cash Crops. Elsevier Science B. V., Amsterdam, pp. 808.

Pan Z., Yang X.-B., Li X., Andrade D., Xue L, and McKinney N., 2010. Prediction of plant diseases through modeling and monitoring airborne pathogen dispersal. *CAB Reviews: Perspectives in Agriculture, Veterinary Science, Nutrition and Natural Resources*, 5, No. 18.

Prévost B., 1807. Memoire sur la cause immediate de la carie ou charbon des bles, et de plusieurs autres maladies des plantes, et sur les preservatifs de la carie. Paris. English translation by Keitt G. M., in *Phytopathological Classics* No. 6, 1939.

Richard J. L., 2007. Some major mycotoxins and their mycotoxicoses - An overview. *International Journal of Food Microbiology*, 119, 3–10.

Russell P. E., 2005. A century of fungicide evolution. *Journal of Agricultural Science*, 143, 11-25.

Shaw M. G., 1994. Modeling stochastic processes in Plant Pathology. *Annual Review of Phytopathology*, 32, 523-544.

Steffens J. J., Pell E. J. and Tien M., 1996. Mechanisms of fungicide resistance in phytopathogenic fungi. *Current Opinion in Biotechnology*, 7, 348-355.

Thind T. S. (ed.), 2012. Fungicide resistance in crop protection. *Risk and management*. CAB International, UK, pp. 284.

Tillet M., 1755. Dissertation sur la cause qui corrompt et noircit les grans de ble dans les epis et sur les moyens de prevenir ces accidents. Bordeaux, pp. 150. English translation by Humphrey H. B., in *Phytopathological Classics*, 1937, No. 5, 1937.

United Nations, Department of Economic and Social Affairs, Population Division, 2011. World Population Prospects: The 2010 Revision, Highlights and Advance Tables. Working Paper No. ESA/P/WP.220.

Van der Plank J. E., 1960. Analysis of epidemics. In: Horsfall J. G. and Cowling E.B. (eds), Plant Pathology, Academic Press, New York, USA, 230-287.

Van der Plank J. E., 1963. Plant diseases: epidemics and control. Academic Press, NY., USA, pp. 349.

Van Maanen A. and Xu X. M., 2003. Modelling plant disease epidemics. *European Journal of Plant Pathology* 109, 669–682.

Vercesi A., Toffolatti S. L., Zocchi G., Guglielmann R. and Ironi L., 2010. A new approach to modeling the dynamics of oospore germination in *Plasmopara viticola, European Journal of Plant Pathology*, 128, 113-126.

Vurro M., Bonciani B. and Vannacci G., 2010. Emerging infectious diseases of crop plants in developing Countries: impact on agriculture and socio-economic consequences. *Food Security*, 2, 113-132.

Wood P. M. and Hollomon D. W., 2003. A critical evaluation of the role of alternative oxidase in the performance of strobilurin and related fungicides acting at the Qo site of complex III. *Pest Management Science*, 59, 499-511.

Zhonghua M. and Michailides T. J, 2005. Advances in understanding molecular mechanisms of fungicide resistance and molecular detection of resistant genotypes in phytopathogenic fungi. *Crop Protection*, 24, 853-863.

In: Fungicides
Editors: M.N. Wheeler, B.R. Johnston

ISBN: 978-1-62948-043-5
© 2013 Nova Science Publishers, Inc.

Chapter 4

CLASSIFICATION OF FUNGICIDES

Erlei Melo Reis[1] and Marcelo Aníbal Carmona[2]
[1]Faculty of Agronomy and Veterinary Medicine, University of Passo
Fundo, Passo Fundo, Brazil
[2]Faculty of Agronomy, University of Buenos Aires, Buenos Aires,
Argentina

ABSTRACT

Fungicides are chemicals agents that inhibit or eliminate the mycelial
growth or fungal spores. The chemical, physical, and biological
characteristics of a fungicide determines its suitability for control of a
determined disease. Fungicides can be classified in different ways as
fungistactic, anti-sporulant, disinfectant, preventative, protectant,
residual, contact, curative, eradicative, topic, deep penetrating, non-
penetrating, mesostemic, and systemic, among others. As a result, the
literature is confusing to clearly understand and to discuss topics on
fungicides. This chapter provides information regarding fungicides
classification covering mainly fungicides used in aerial organs of plants.

1. INTRODUCTION

Fungicides have been classified in as many different ways as fungistactic,
anti-sporulant, disinfectant, preventative, protectant, residual, contact,
curative, eradicative, topical, deep penetrating, non-penetrating, mesostemic,
and systemic, among others. As a result, the literature is confusing. In

addition, fungicides can be classified according to at least six aspects: according to where they are applied, according to the mobility and/or position in the plant, according to the time of application and sub-phases of the infectious process, according to the fungicide uptake by the fungus and according to their mechanism of action.

This classification covers mainly fungicides used in aerial organs of plants, and a given fungicide may be categorized in one or more of the different aspects (Reis et al. 2010).

2. CLASSIFICATION OF FUNGICIDES ACCORDING TO WHERE THEY ARE APPLIED

Fungicides are used to control fungal pathogens present (a) in the soil, (b) in seeds, (c) in above-ground organs: (c1) in resting fruits in winter, (c2) on green organs, (d) in post-harvest treatment and (e) in wood preservation. In each of these cases, fungicides present different characteristics.

2.1. Fungicides Applied to the Soil — Disinfectant Fungicides

The soil is a complex and dynamic environment due, among others, to the microbial activity, fungicide fixation on clays and in organic matter, hydrolysis and pH of the soil solution, all of which contribute to the reduction of the useful life of the applied chemical. Therefore, fungicides applied to the soil must be stable. Methyl bromide and formaldehyde are used exclusively in the soil and are considered contact biocides.

2.2. Fungicides Applied on Seeds - Seed Dressing Fungicides

Protectant fungicides applied on seeds remain on their surface, forming a protectant layer on the seed surface, able to protect them from soilborne fungi and prevent fungi associated to seeds (as internal mycelium) from reaching the seed surface and colonizing aerial organs and roots. If fungi reach the coleoptile and the plumule, they can colonize them, reach the soil surface and then be disseminated by the secondary cycles to the foliage. Protectant fungicides prevent the fungal transmission or passage from the inner endosperm to aerial organs, a process which does not require spore germination. These fungicides should be directly absorbed by the mycelium when reaching the seed surface after water embebition in the soil. When the seed absorbs water from the soil, through the process of soaking and germination, the pathogen also resumes its vital activities, seeking to grow from inside the seed to the aerial organs, which will fulfill its vital function.

The so-called systemic fungicides applied to seeds are not absorbed or translocated, because seeds have no vascular system. These fungicides remain on the surface and are then leached into the soil. During and after seed germination, they are up-taken by the seminal roots and translocated via the xylem to the seedling aerial organs. Fungicides that control the attack by *Blumeria graminis* (powdery mildew of oats, barley and wheat) act via the xylem at the time spores are deposited, germinate and penetrate the protected leaves. In the control of rusts in winter cereals, the process is also via root uptake and translocation via the xylem. The prolonged protection, up to 60 days, is due to the chemical slow release from the seed surface and absorption via roots.

Captan, carbendazim, carboxin, difenoconazole, iprodione, fludioxonil, metalaxyl, thiram, tolylfluanid and triadimenol are examples of seed-dressing fungicides.

2.3. Fungicides Applied on Leaves - Deposited by Spraying

2.3.1. *Winter Treatment of Deciduous Leaves Fruit Trees*

The fungicides used in winter treatment have contact action on the direct cell wall penetration of fungi. Contact fungicides are those that aim to reach the fungus in its resting phase, both before and after reaching the infection site. These fungicides are applied when spores, or the inoculum, are present on the plant surface in the dormancy season of fruit species because the absence of leaves allows the application of phytotoxic fungicide.

When contacting any kind of fungal inoculum (active spores, dormant spores, mycelium and resting structures), the fungicide penetrates and kills the fungus, not requiring spore germination (Horsfall, 1957). Contact fungicides used in winter treatment include lime sulfur, Bordeaux mixture and copper-based products in large doses.

2.3.2. Treatment of Green Aerial Organs in the Growing Season

The use of resistant genotypes is the most effective and economic strategy to prevent diseases. However, the level of resistance present in the cultivated germplasm of most crops is not sufficient. In this way, the disease intensity may be suppressed or prevented by foliar application of fungicides. For this reason, fungicides are used primarily to treat green organs of cultivated plants. Most of the crops are protected by different fungicides or their mixtures. Besides, several different chemicals may be used for the same crop or disease.

2.3.3. Fungicides in Post-Harvest Treatment

The objective of fungicides in post-harvest application is to protect yield quality and maintain the value of the production from storage pathogens. Several fungicides are used in fruit, floral, and horticultural crops after harvest.

2.3.4. Fungicides in Wood Preservation

Wood is a natural organic material and as such it can be degraded by fungi. Wood preservative fungicides are a very diverse organic and inorganic group of chemicals, commonly used for preservation of wood against, fungi. The chemical and biological characteristics of a wood fungicide determine its suitability and efficacy.

3. CLASSIFICATION OF FUNGICIDES ACCORDING TO THE MOBILITY AND/OR POSITION IN THE PLANT

Fungicides can be classified primarily in relation to their position in the plant: whether they remain on the surface after deposition or are absorbed and transferred by the vascular system to distant sites from deposition. The uptake and mobility of the fungicide in the plant will depend on the attributes of the chemical molecule.

a. Non-Penetrating Fungicides - Topical, Non-Mobile

The term penetration indicates that the action is tied to the molecule attributes and that the absorption/adsorption action is governed by the plant cell.

Non-penetrating fungicides applied to the aerial organs are not absorbed and therefore not translocated, remaining on the plant surface on the site (Greek, topykos = place) where they were deposited. They remain on the leaf surface, acting as a shield, and only provide a barrier to protect against infection. Therefore, they can be washed off from the plant surface by rain water.

Non-penetrating fungicides include captan, chlorothalonil, cupric salts, dithianon, famoxadone, fluazinam, folpet, iprodione, maneb, mancozeb, pencycuron, prochloraz, propineb, quinomethionate, sulfur, thiram, triphenyl tin acetate, triphenyl tin hydroxide, vinclozolin and zoxamide.

Some references consider that topical fungicides (non-penetrating) as synonymous of contact fungicides (FRAC - UK, 2007). However, this does not look completely right as regards the meaning of both terms.

b. Penetrating- Mobile, Inside the Plant

Penetrating fungicides are deposited on the surface of the susceptible organ and move (penetrate, are absorbed) to the inside of leaf tissues. The active ingredient penetrates the plant tissue, has laminar activity, but there is little or no transport within the vascular system of the plant.

Locally Penetrating Fungicides

Movement of Depth - Partially Vertical Movement

The depth action occurs when the fungicide is applied on the leaf surface and is translocated to the inside of the leaf. Examples include strobilurins and triazoles.

Locally penetrating fungicides are absorbed into the immediate area of application but are not translocated far from the site of uptake. They serve to prevent the development of disease at (and in a small zone surrounding) the site of uptake, and do not pass from one epidermis to the other, thus are not translocated via vessels. The movement is vertical. An example includes dodine applied on apple leaves.

Translaminar Movement - Partially Mobile in Vertical Direction

When the absorbed fungicide passes from one leaf face to the other but is not transported to locations distant from the deposition site, it is called fungicide with translaminar movement. These fungicides are not truly systemic and do not move throughout the plant. The movement of these fungicides is vertical, but longer than of local penetrating fungicides. An example of this kind of fungicide includes triazoles in soybean leaflets.

Local Systemic Movement - Partially Mobile in Lateral and Vertical Direction

When the fungicide penetrates and is translocated to a short distance near the site of deposition it is called local systemic fungicide. The movement is lateral and vertical. The active ingredient reaches areas on the same leaf that were not directly sprayed. An example includes triazoles in broad-leaved crops.

Mesostemic movement. Mesostemic fungicides are strongly attracted to the leaf surface by providing a barrier against disease infection. Small amounts penetrate the leaf and move to the opposite, untreated surface of the leaf, providing protection against infection on both leaf surfaces. Mesostemic fungicides prevent spore germination, meaning that the pathogen will not penetrate the host plant. The active ingredient reaches areas on the same leaf that were not directly sprayed. The term mesostemic was coined to classify strobilurins. A chemical is present when there is mesostemic affinity with the leaf surface and when it can be absorbed by the wax layer, forming a deposit on the surface of the susceptible organ. Subsequently, the product can be redistributed to the surface of the plant in its vapor phase. The substance penetrates the tissues presenting mesostemic translaminar activity, but with minimal or nonexistent vascular translocation (via the phloem or the xylem). Mesostemic fungicides are used based on the site where they act, i.e. the mesophyll. The mesophyll is the fundamental leaf industry, which is located between the upper and lower epidermis, specializing in photosynthesis. Mesostemic fungicides are deep-penetrating with translaminar movement or action. Fungicides with this property are not systemic, but form:

a)   a free deposit which can be redistributed by water;
b)   a more cohesive weather-resistant deposit on the leaf surface;
c)   a deposit strongly associated with the cuticular wax layer very resistant to removal by rain or bleach, allowing a long residual effect; redistribution on the leaf surface occurs through absorption from the continuous layer of cuticular wax of the leaves to the inside organ and also through the vapor phase and reabsorption by cuticular wax; this feature confers protective action
d)   a fraction that penetrates the leaf tissue.

Therefore, mesostemic fungicides have lypophylic characteristics, and their deposits adhere strongly to the wax layer of the cuticle and have greater resistance to removal by rain or irrigation. The main mesostemic fungicides are the strobilurins azoxystrobin, kresoxim methyl, picoxystrobin, pyraclostrobin and trifloxystrobin (FRAC Code List 2, 2012).

Systemic Movement - Penetrating Mobile Molecule

Systemic fungicides are those absorbed by the roots or leaves and subsequently translocated through the plant via the xylem and phloem.

Translocation is the movement of the chemical within the plant body to tissues distant from the deposition site. The movement via the xylem, or acropetal movement, is the most common, as with benzimidazoles and triazoles. Triazoles translocate mainly via the xylem, but also translocate partially via the phloem.

The movement via the phloem, or basipetal movement, is more difficult. Only fosetyl aluminum presents this property and is thus considered a true systemic fungicide. Therefore, systemic fungicides differ from topical fungicides (non-penetrating) by their ability to be redistributed within the treated organs, especially leaves.

The term "systemic" may lead to the idea that the fungicide with this property can be transported to any plant part, for example deposited on a leaf and then moving to another leaf or to the upper or lower canopy and vice versa. Inside a gramineous leaf, translocation is complete, from the base to the apex, but not from the apex to the base. Moreover, many fungicides that are mobile in broad-leaved plants have only a local systemic action or are translocated only to short distances within the leaf from the point of deposition, thus needing good coverage to obtain maximum control efficiency.

The search for chemical synthesis of new fungicides focuses on developing systemic fungicides with acro- and basipetal translocation. Once inside the plant, systemic fungicides confer a protective action longer than that of residual fungicides (15 to 20 days), not exposed to wash and photo-decomposition, thus requiring fewer and less frequent applications. It has also been observed that, after application, systemic triazoles in winter cereals the occurrence of rain one or two hours after spraying does not compromise negatively the control.

Fungicides absorbed via roots protect the entire aerial part of the seedling, differently than when applied to the foliage. Fosetyl aluminum, sprayed on the foliage of citrus and apples, is absorbed and translocated subsequently to the root via the phloem, where it controls fungi of the genus *Phytophthora*, the causal agent of root rot. Therefore, it is cited as a true systemic fungicide.

The main penetrating-mobile fungicides are bitertanol, carbendazim, carboxin, kasugamycin, cymoxanil, cyproconazole, difenoconazole, dimethomorph, epoxiconazole, fenarimol, fenamidone, fluquinconazole, flutriafol, fosetyl aluminum, hexaconazole, iprovalicarb, metalaxyl, metconazole, myclobutanil, procymidone, propamocarb, propiconazole, prothioconazole, tebuconazole, tetraconazole, thiabendazole, thiophanate methyl, triadimenol, and tricyclazole triflumizole. Azoxistrobin in the only mobile strobilurin.

4. CLASSIFICATION OF FUNGICIDES ACCORDING TO THE TIME OF APPLICATION AND SUB-PHASES OF THE INFECTIOUS PROCESS

In this classification criterion, the action of the fungicide is directly related to the fungus (fungicide x fungus relationship). Non-penetrating fungicides act before penetration of the fungus, on the plant surface, which depends on the time of application. Moreover, fungicides can have penetrating effect on the fungus when they are within the host tissues (penetrating/colonization sub-phases), also dependent on the time of application.

According to Hewitt (1998), regarding the time of application, fungicides can be classified as preventive, curative and eradicant, according to the infection sub-phases wherein the fungicide operates

The infectious process comprises the following sub-phases: inoculum deposition at the infection sites, spore germination, germ tube penetration and the beginning of host colonization. Colonization means the invasion and nutrient extraction from host tissues with subsequent manifestation of symptoms/signs.

a. Protectant – Non-Penetrating, Fungicide Action on the Plant Surface

The terms "preventive" and "protective" have been used interchangeably. "Preventive" means that it prevents, whereas "protectant" means that it protects (in this case from infection). The term "residual" comes from the deposit, a property of non-penetrating fungicides.

These fungicides fall into non-penetrating fungicides, in which a deposit remains on the surface of the organ treated even after harvest. In the latter case, the deposit is called residue.

Preventive or protectant fungicides are applied before spores are deposited at the host infection sites. The action is protectant or of pre-penetration. Before the inoculum is deposited, the fungicide forms a layer on the surface of the organ, is up-taken by the fungus pro-mycelium during germination and preventing penetration.

In practice, under field conditions, the time of spore deposition at the infection sites is not known. When no symptoms/signs are displayed, one admits that the treatment was performed preventively.

b. Curative – Penetrating Fungicide Action within the Plant

When a fungicide application interrupts the development of an established infection, i.e. the plant shows no visible disease symptoms, its action is called curative.

The curative action occurs after the penetration of the fungus into the plant, but while no symptoms/signs are yet observed (pre-symptom/sign period).

The visual effects are similar to those produced by protectant topical fungicides, i.e. absence of symptoms. The curative action occurs when penetrating fungicides come into contact with the fungal structures within the plant tissues. Therefore, no non-penetrating fungicides have curative action, since they act on the surface of the treated plant.

During the curative action, both the pathogen structures and the fungicide are within the plant tissue. To be curative, products must be up-taken by plant tissues. If the fungicide is effectively curative, latent infections are controlled and the infectious process is paralyzed. This action occurs, for example, for DMI fungicides (demethylation inhibitors), which penetrate soybean leaves and thus control *Phakopsora pachyrhizi*.

In the case of wheat leaf spot, caused by *Bipolaris sorokiniana*, the curative action increases with the time elapsed after the fungicide application, and achieves fungal eradication 15 days after application. Differences between DMIs applied alone and when applied in mixture with strobilurin can be observed .

Based on the meaning proposed by Hewitt (1998), the term curative should not be used as a decision-making criterion for the first application of fungicides in farming. The curative action has no practical value because it is not possible to be detected and quantified in the field.

c. Eradicant Fungicides - Penetrating Fungicide Action within the Plant

In this case, the eradicant fungicidal activity involves the effects of chemicals on the post-symptom/signal stage i.e. when the application of the fungicide interrupts the further development of an established infection with visible disease symptoms on the plant. In any curative or eradicant case, the action refers only to the death of the fungus, without regeneration of host cells or dead tissues. Death of cells and tissues in plants is generally irreversible.

For leaf spots in winter cereals, some fungicides can halt the lesion expansion with the death of the pathogen involved. Zanatta (2007, unpublished data) observed that the death of the fungus occurred 15 days after application of the fungicide, but that the symptoms did not disappear.

Most DMI fungicides present eradicant action against fungi of the genera *Blumeria, Erysiphe, Phakopsora, Puccinia, Uromyces* (in aerial organs) and *Ustilago* (which infects the embryo of winter cereal the seed). These fungicides kill pathogens, but the injured tissues, especially leaves, do not regenerate.

In a few cases, non-penetrating and penetrating fungicides may have curative and eradicative action, although this is more common for penetrating fungicides. For example, after penetrating the host, powdery mildews (order Erysiphales) develop externally on the plant surface, and are therefore controllable by non-penetrating fungicides such as sulfur-based ones. This, however, is an exception.

Due to its eradicant action, fosetyl aluminum paralyzes the parasitic phase (colonization) of fungi that cause root rot of citrus and apples (caused by *Phytophthora* spp.) by activating the plant self-defense mechanism.

Evans (1973) and other authors considered that the terms "curative" and "eradicant" are synonymous while Hewitt (1998) separates them.

5. CLASSIFICATION OF FUNGICIDES ACCORDING TO THE FUNGICIDE UPTAKE BY THE FUNGUS

Fungicides can be classified into residual and contact according to their absorption by the fungal structures (for example by spores). Absorption occurs on the surface of the susceptible organs, usually leaves. The concept of contact fungicides is not sufficiently clear in the literature. For many authors, it is synonymous of topical and non-penetrating. However, in his book "Principles of fungicide action", Horsfall (1957) classifies protectant fungicides into two groups: contact and residual. This classification refers to the fact that germination of the propagule may be either required or not to be absorbed by the fungal spores.

If the propagule is active (i.e., it germinates), fungicides can be protective (pre-penetration), curative (post-penetration, pre-symptom/sign) or eradicant (post-symptom/sign). Moreover, for the dormant pathogen, the contact fungicide has to be absorbed directly by the fungal cell wall.

If the infection does not become established, a fungicide will be contact or protective. If the spore is not dormant and germinates, the fungicide prevents penetration and has a protective or residual action.

a. Contact – Penetrating Fungicide in the Dormant Fungal Cell Wall

Contact fungicides are those aimed to reach the fungus in its resting phase, both before and after the fungus have found the infecting site. The compound is applied, for example, when spores or the inoculum are present on the surface of fruit trees in the dormant season which come to rest in winter. When contacting any fungal structure (spores, dormant spores, mycelium), the fungicide penetrates it and kills it, without requiring germination (Horsfall, 1957). Contact action is an attribute of the fungicide.

Due to its characteristics, contact fungicides are used in the winter treatment of fruit trees. Examples of contact fungicides are lime sulfur, the Bordeaux mixture and copper-based products applied at higher doses.

It is not fully clear whether strobilurin and triazole have contact action. If applied directly to the spores of the fungi that cause rust (e.g. leaf wheat and soybean) and powdery mildew (in winter cereals and soybean), these fungicides could determine their death. In this case, the fungicidal action would occur by direct penetration of the fungicide through the cell wall (mycelium and spores) without the need of spore germination.

The term "contact" has been incorrectly used as a synonym of protectant or topical (non-penetrating). In addition, it should not be used as a synonym of eradicant. It is proposed that the word "eradicant" should be used in the sense defined by Hewitt (1988).

According to several authors and FRAC-UK (http:frac.csl.gov.uk), a contact fungicide is one that acts directly in contact with the "disease" and does not move in the plant. This term may be confused with topical or non-mobile, or non-penetrating. It must be remembered that topical fungicides are not mobile.

b. Protectant or Topical Fungicides – Protection Afforded by the Deposit of the Fungicide on the Leaf Surface

Protectant fungicides are topical and act on the plant surface. Topical protectant fungicides are applied in a prior toxic layer on the host surface and "wait" for the mobile propagule, i.e. the spores, before penetration. When spores germinate, the toxic penetrates and kills the germling, thus preventing penetration. The immature germ tube and appressorium are sensitive to residual fungicides. The appressorium imbibes water to create the turgor pressure required to penetrate the leaf tissue. By soaking, it absorbs water along with the diluted fungicide.

The term protective fungicides must be applied to non-penetrating or topical fungicides, being understood that the deposit of the fungicide lies on the plant surface after evaporation of the applying vehicle, i.e. water.

This group includes fungicides used to control diseases in McNew's (1962) Group V, which interfere with the photosynthetic process. Protective fungicides are applied to aerial organs of plants, forming a protective toxic layer on the surface (non-penetrating). Thus, when the inoculum (spores) is deposited on susceptible tissues and germinates, the toxic in contact with the germ tube penetrates it and, by biochemical mechanisms, thus determines the protoplasm death.

The topical protective action aims to prevent penetration, thus preventing the infection that would occur in the future. Therefore, topical protectant fungicides act by preventing or reducing the penetration of host tissues and reducing the number of penetrations or future lesions.

Topical fungicides are deposited on the plant surface and can suffer redistribution by rain or dew. It has been reported that a rainfall of 13 mm removes the deposit of this type of fungicide in tomato, and that one of 23 mm does so in apple.

Penetrant fungicides, absorbed by leaves, can paralyze the penetration process, thus preventing parasitism. This occurs when the penetration tube, after going through the cuticle, reaches the epidermal/mesophyll cells containing inhibitory concentrations of the fungicide to the invader fungus. The parasite dies during penetration when it absorbs the toxic before the establishment of parasitism. According to Hewitt (1998), this characterizes the curative action that occurs within the leaf. In the case of DMIs, fungicidal action occurs 24 hours after initiation of spore germination when ergosterol synthesis starts.

The duration of the protection period conferred by chlorothalonil is longer than that conferred by mancozeb, whose deposit remains for more time on the treated surface.

After penetration, when the pathogen is inside the plant, protectant, non-penetrating fungicides have no capability to prevent the further fungal invasion, i.e. have no curative or eradicant action.

The main non-penetrant fungicides applied in above-ground plant parts are captan, cartap (insecticide with a fungicide), chlorothalonil, copper-based products (copper hydroxide, basic copper sulphate, cuprous oxide and copper oxychloride), dithianon, dodine, wettable sulfur, fluazinam, folpet, iprodione, maneb, mancozeb, oxadixyl, pencycuron, prochloraz, propineb, quinomethionate drawing, tolylfluanid, vinclozolin and zoxamide. These fungicides must be applied before spore deposition on the infection sites.

The main seed protectant fungicides are captan, carboxin iprodione, quintozene and thiram.

6. CLASSIFICATION OF FUNGICIDES ACCORDING TO THEIR MECHANISM OF ACTION

The mode of action shows how a fungicide kills or suppresses a target fungus, i.e. the specific biochemical process of the target fungus affected by a fungicide. Examples are damaging cell membranes, inactivating critical enzymes or proteins, or interfering with key processes such as energy production or respiration.

After being absorbed by the fungal cells, fungicides act in some cellular biochemical functions resulting in fungal death. For details on the mode of action of fungicides, read FRAC Code List (2012).

REFERENCES

Frac, 2012. Fungicide resistance action committee. Global Crop Protection Organization. Brussels. *(www.gcpt.org/frac)*.

Hewitt, H. G. *Fungicides in crop protection*. CAB International, 1998. Chapter 4. Fungicide Performance. p. 87-153.

Horsfall, J. G. *Principles of fungicidal action*. Walthan, Mass: Chronica Botanica, 1957. 279p.

Mcnew, G.L. The nature, origin, and evolution of parasitism. In: HORSFALL, J.G. & DIMOND, A.E., eds. *Plant pathology*, New York, Academic Press, v.2, p.2-66. 1960.

Nene, Y. L.; Thapliyal, P. N. *Fungicides in plant disease control*. 3. ed. New York: International Scientific Publisher, 1971. 507p.

Reis, E. M.; REIS, A. C.; CARMONA, M. A. *Manual de fungicidas – Guia para o Controle Químico de Doenças de plantas*. Passo Fundo, RS, Brasil, Editora UPF.2010,

Zanatta, T. Efeito da aplicação de fungicidas no processo infeccioso de *Phakopsora pachyrhizi*, em soja. *Dissertação de mestrão*. Universidade de Passo Fundo. 2009.

In: Fungicides   ISBN: 978-1-62948-043-5
Editors: M.N. Wheeler, B.R. Johnston   © 2013 Nova Science Publishers, Inc.

Chapter 5

EFFECTIVE FUNGICIDES FOR CEREAL CROPS PROTECTION AGAINST TOXICOGENIC FUNGI CAUSING *FUSARIUM* HEAD BLIGHT

L. D. Grishechkina and V. I. Dolzhenko

All Russia Plant Protection Research Institute, St.-Petersburg, Russia

ABSTRACT

Analyses the long-term effectiveness studies of novel fungicides use for *Fusarium* head blight control on winter wheat in Krasnodar territory of Russia are presented. Effects of fungicides on the grain infestation by the pathogen including latent infection are described. The hazard of studied preparations to the grain agrobiocenosis is shown.

Keywords: Winter wheat, fungicides, *Fusarium* head blight

BACKGROUND

The majority of modern fungicides used in agriculture for *Fusaruim* species control are not always active against causative agent of cereals *Fusarium* head blight. The efficiency of widely used fungicides based on benzimidazole (benomyl, carbendazim), triazole and imidazoles (prochloraz) etc. does not exceed 40–60 % and only for tebuconazole it reaches 70–80 %. High fungicide activities against causative agents of fusariose show methconazole and prothioconazole. Modern preparations on cereals suppress the wide micromycetes' complex development but some of them are not effective against fusarioses, in particular propiconazole. As a result of its application the disease development was limited only onto 20–30 % (Zazimko et al., 1992) and *Fusaruim* infection accumulation in spiked cereals agrobiocenosis of industrial wheat crops in Krasnodar territory of Russia was promoted.

There is no agreement of opinion concerning the strobilurins' efficiency. A number of researchers (Grossman, Retzlaft, 1997; Greenfield, Rosall, 2000) consider strobilurins also do not affect pathogen, enlarging the mycotoxins' content in grain. In Russia strobilurins' application is regulated, they are used basically in fit combinations. Last years the fungicides assortment was filled up by the combined preparations containing active substances with various fungicide activity and action spectrum. It demands their activity against pathogen studying.

On the application efficacy of plant-protecting agents a number of reasons connected with pathogeny affects: polycyclic history; expanded infestation of plants; penetration speed of infecting and distributing into internal plant tissues within the spike limits, caryopsides' colonization degree, disease latent period, squames bedding to caryopsis intensity; rates of grain forming and maturing etc. At the same time temperature, humidity, plants' disease resistance, fungicide dosages; time of spraying, pathogen response to fungicide, quality of spike covering with fungicide and some other reasons affect the micromycete behavior and fungicides' application efficiency.

The maximal fungicides' application effect is provided with optimum treatment terms of vegetans: for barley with flowering enclosed-type is heading (Boiko, Buga, 2003), for wheat is 2–4 days before flowering: the heading end – the start flowering. M. I. Zazimko (2012) guess that the most vulnerable phase for pathogen in Krasnodar territory is heading, the other researchers (Monastyrnaya, Androsova, 1992) subsume this period to wheat full flowering phase.

In favorable conditions for disease development (crop rotation ignoring, tilling infringement, corn crops neighborhood to spiked cereals against resistant varieties' insufficient assortment) plants' treatment with chemicals is compulsory. The pesticide selecting will determine its efficiency in spike disease distribution decreasing, grain infestation reducing; productivity increasing; mycotoxins' level in grain decreasing. In fungicides' efficacy estimating the great value is given also to ecological hazard of their application. Such researches should take into account a degree of pesticide hazard and speed of their degradation in environment up to nontoxic products because there is too short period after chemical treatment before grain maturing.

The basic purposes of our researches were the comparative biological and economic efficiency estimations of modern fungicides against *Fusarium* head blight on winter wheat variety Bat'ko as well as their ecological estimation for cereals agrobiocenosis. In our work the preparations have been used: Folicur, EC (emulsifiable concentrate) (250 g/l tebuconazole) at the application rate 1.0 l/ha; Forus, EC (125 g/l tebuconazole + 100 g/l triadimedfon) – 1.0 l/ha; Prosaro, EC (125 g/l prothioconazole + 125 g/l tebuconazole) – 1.0 l/ha; Zamir *, EW (oil in water emulsion) (133 g/l tebuconazole + 267 g/l prochloraz) – 1.2 l/ha; Titul Duo, CSC (colloid suspension concentrate) (200 g/l tebuconazole + 200 g/l propiconazole) – 0.32 l/ha; Falcon, EC (250 g/l spiroxamine + 167 g/l tebuconazole + 43 g/l triadimedfon) – 0.6 l/ha; Impact, SC (suspension concentrate) (250 g/l flutriafol) – 0.5 l/ha; Acanto Plus, EC (200 g/l picoxystrobin + 200 g/l ciproxonazole) – 0.6 l/ha; Alto Super, EC (250 g/l propiconazole + 80 g/l ciproxonazole) – 0.5 l/ha; Amistar Trio, EC (125 g/l propiconazole + 100 g/l azoxystrobin +30 g/l ciproxonazole) – 1.0 l/ha. As ecological safety criterions the integrated indices served for investigated protective means – toxic load and persistence index.

METHOD

Researches were carried out on winter wheat variety Bat'ko in Krasnodar territory according to methodical instructions (2009). The size of the experiment plots was 10 m^2 and replication was fourfold. Agricultural practices on the experimental plots were the conventional ones for the given region. The soil was very deep leached chernozem with 3.8 % humus content, pH = 6.7. Infection background was prepared by wheat infestation with water conidial suspension of *Fusarium graminearum* in concentration × 10^5 CFU (colony forming units)/ml. Suspension applied on spike in the start flowering stage at rate 50 ml suspension/m^2. Then the inoculated plants were covered with polyethylene isolators, making the moist chamber within 24 hours. Further they were treated with fungicides according to the experimental variants with rate of 300 l/ha working liquid by "Tecnoma" brand sprayer. In control variant plants inoculated with the disease pathogen were not treated with fungicides. Under field conditions head blight infection was scored visually with a help of 5-grade scale. Grain infestation was determined by counting seeds with typical disease signs, and latent head blight infection by caryopsides' phytopathological analysis.

The pesticide's ecological hazard index was determined based on toxic load (Fadeev, 1988). The hazard class and persistence index of active ingredients in plant tissues (Burov et al., 1995) worked out with a help of 4-grade scale. Pesticide active ingredients' residues in winter wheat tissues in dynamics were managed by high performance liquid and gas-liquid chromatography methods. Statistical processing of the data was carried out by the ANOVA method (Dospekhov, 1973).

RESULTS AND DISCUSSION

Fungicide treatments of wheat engaged for *Fusarium* head blight reducing onto 49.0–77.0 % depending on the agroecological conditions of the test years at head blight affection in the control up to 69.7–78.6 % (fig. 1). The greatest efficiency in disease managing was provided by Zamir, EW. Folicur, Falcon and Acanto Plus efficacy kept pace with 63–65 %.

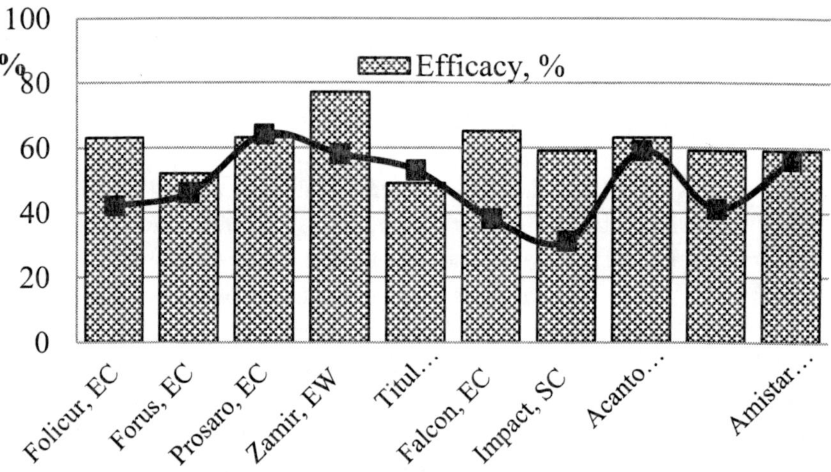

Figure 1. Modern fungicides' biological and economic efficiency in winter wheat *Fusarium* head blight control (variety Bat'ko), years 2008–2010.

Table 1. Fusarium head blight infection of winter wheat grain (variety Bat'ko)

| Active ingredient | Commercial name of chemicals | Application rate, l/ha | % grains | | Grain weight, g |
			with *Fusarium* head blight symptoms	without *Fusarium* head blight symptoms	with *Fusarium* head blight symptoms
Tebuconazole	Folicur, EC (250 g/l)	1.0	43.3	56.7	36.9
Tebuconazole + triadimedfon	Forus, EC (125 + 100 g/l)	1.25	48.0	52.0	36.2
Prothioconazole + tebuconazole	Prosaro, EC (125 + 125 g/l)	1.0	49.2	50.8	31.0
Tebuconazole + prochloraz	Zamir, EW (133 + 267 g/l)	1.2	28.3	71.7	29.8
Tebuconazole + propiconazole	Titul Duo, CSC (200 + 200 g/l)	0.32	61.3	38.7	42.0
Spiroxamine + tebuconazole + triadimedfon	Falcon, EC (250 + 167+43 g/l)	0.6	25.2	74.8	41.4
Flutriafol	Impact, SC (250 g/l)	0.5	50.0	50.0	13.2
Picoxystrobin + ciproxonazole	Acanto Plus, EC (200 + 200 g/l)	0.6	39.7	60.3	36.6
Propiconazole + ciproxonazole	Alto Super, EC (250 + 80 g/l)	0.5	42.9	57.1	38.2
Propiconazole + azoxystrobin + ciproxonazole	Amistar Trio, EC (125 + 100 + 30 g/l)	1.0	48.4	51.6	41.2
Control (without treatment)	-	-	74.4	25.6	74.6

During all the research years, even under the extreme weather conditions during the year 2010 with prevalence of elevated air temperatures and rain precipitation absence chemically treated plants provided guaranteed yield increasing up to 31–84 % in comparison with control plants. 1000 grains weight marker in variants of chemicals' application was higher than the control one over 5.2–19.0 g. In all variants of fungicides' application caryopsides' fusariose infestation was reduced. The best results were obtained by Zamir, EW and Falcon, EC use where caryopsides' fusariose infestation reached only

25.2–28.3 %, and in the control was up to 74.4 % (tab. 1). Fungicide affect was distributed on disease latent form therefore micromycete infection load after spraying has decreased especially by Zamir, EW; Falcon, EC; Folicur, EC, Acanto plus, EC and Amistar Trio, EC etc. application.

All investigated fungicides belong to low hazard compounds, toxic load does not exceed 25.4–123.9 semilethal doses/ha. Among them the least negative effect on the cereal agrobiocenosis act Amistar Extra, Titul Duo, Amistar Trio, Acanto Plus, their toxic load is not more than 25.4–39.9 semilethal doses/ha (tab. 2). We found that some working substances (tebuconazole, prochloraz, triadimedfon, propiconazole etc.) are the persistent agents. Process of their degradation in plant tissues occurs less than 40 days, during the analysis the preparations' residues in grain and straw are not revealed.

Table 2. Fungicides toxic load for wheat Fusarium head blight control

Active ingredient	Active ingredient average application rate, l/ha	Rat LD_{50} mg of active ingredient/kg	Toxic load (number of semilethal doses/ha)
Tebuconazole	0.25	2850	87.7
Flutriafol	0.125	1140	109.6
Picoxystrobin + ciproxonazole	0.24	6020	39.9
Propiconazole + ciproxonazole	0.165	3426	48.2
Propiconazole + azoxystrobin + ciproxonazole	0.255	8426	30.3
Tebuconazole + triadimedfon	0.28	3850	72.7
Prothioconazole + tebuconazole	0.23	9050	25.4
Tebuconazole + prochloraz	0.48	3873	123.9
Tebuconazole + propiconazole	0.128	4367	29.3
Spiroxamine + tebuconazole + triadimedfon	0.276	4145	66.6

Thus, the carried out researches show that modern fungicides essentially limit the head blight development, including caryopsides' latent form infestation. All investigated fungicides pose low hazard for cereals agrobiocenoses because they are less hazardous compounds, their toxic load does not exceed 123.9 semilethal doses/ha.

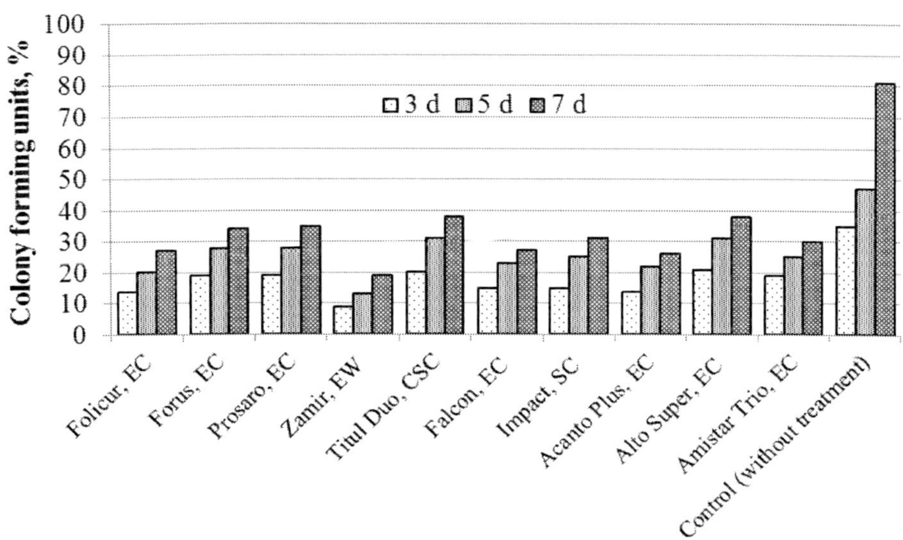

Figure 2. Fungicides' action on the *Fusarium* head blight latent form.

REFERENCES

Burov V. N., Tjuterev S. L., Sukhoruchenko G. I., Petrova T. M. Estimation method of pesticides' ecological safety at their use in the integrated protection of plants. //*Methodical instructions*. SPb. 1995. 14 p.

Dospekhov B. A. *Experimental affair techniques*. - M.: Kolos, 1973. 336 p.

Fadeev Ju. N. Estimation of sanitary and ecological safety of pesticides. // *Protection of plants*. 1988. № 7. p. 20–21.

Greenfield J. E., Rosall S. The effect of fungicides on productivity and toxin content in wheat inoculated with Fusarium. //*6-th European Fusarium Seminar*. Book of abstracts. 2000. p. 104–105.

Grossman K., Retzlaft G. Bioregulatory effect of the fungicidal strobilurin kresoxim-methyl in wheat. //*Pest Sc*. 50. 1997. p. 11–20.

Methodical instructions on fungicides' registration tests in agriculture. SPb. VIZR. 2009. 378 p.

Monastyrnaya E. I., Androsova V. M. Optimum term for fungicides application against Fusarium head blight on winter wheat crops. // Abstracts. Scientific-coordination meeting *"Fusarium head blight on cereals".* Krasnodar. 1992. c. 32.

Zazimko M. I. Kuban experience is useful to everybody. // the Newspaper *"Protection Argument".* 2012. № 5 (36). p. 4–5.

Zazimko M. I., Gonik A. G., Bolbat A. V., Handoga N. V. Fusarium head blight on cereals. // Abstracts. Scientific-coordination meeting *"Fusarium head blight on cereals".* Krasnodar. 1992. c. 31.

INDEX

X

Y

Z